Student's Guide to Writing Dissertations and Theses in Tourism Studies and Related Disciplines

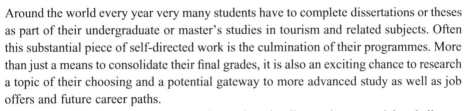

Around the world every year very many students have to complete dissertations or theses as part of their undergraduate or master's studies in tourism and related subjects. Often this substantial piece of self-directed work is the culmination of their programmes. More than just a means to consolidate their final grades, it is also an exciting chance to research a topic of their choosing and a potential gateway to more advanced study as well as job offers and future career paths.

Yet for all these reasons, many students view the dissertation as a tricky challenge. This comprehensive book intends to take the stress and anxiety out of doing a dissertation in tourism studies and related disciplines. The process is examined from the germination of an idea to the submission and assessment of the final document. Written primarily for students conducting independent research for the first time, this book offers simple advice and a clear framework which students can adopt even in more advanced studies at master's and doctoral level. This book debunks popular myths, and aims to overcome common pitfalls. It focuses on the aims and objectives as the DNA of every dissertation. Rather than view it as a single, overwhelming project, the dissertation is presented as a series of more modest, manageable yet crucially interlinked tasks that all students can successfully complete through careful preparation and effective time management.

Dissertations are not to be underestimated and they demand great care and attention, but they can also be immensely rewarding and enriching experiences academically and personally. This 'jargon free' book is also written with overseas students specifically in mind, drawing directly on our overseas students' experiences. This valuable resource contains start of chapter learning objectives and end of chapter checklists, as well as numerous boxed case studies, to further help assist students through their dissertations.

Tim Coles is Professor of Travel and Tourism Management at the University of Exeter Business School, UK, and Director of the ESRC-funded Centre for Sport, Leisure and Tourism Research, an internationally recognized centre of excellence in postgraduate research training.

David Timothy Duval is Associate Professor in the Faculty of Business and Economics at the University of Winnipeg, Canada. He is also Honorary Associate Professor in the School of Business at the University of Otago and a Senior Visiting Fellow in the School of Aviation at the University of New South Wales, Australia.

Gareth Shaw is Associate Dean for Research and Knowledge Transfer at the University of Exeter Business School, UK. As Professor of Retail and Tourism Management, over 50 of his students have completed PhDs.

Student's Guide to Writing Dissertations and Theses in Tourism Studies and Related Disciplines

Tim Coles, David Timothy Duval
and Gareth Shaw

Routledge
Taylor & Francis Group

NEW YORK AND LONDON

First published 2013
by Routledge
2 Park Square, Milton Park, Abingdon, Oxon OX14 4RN

Simultaneously published in the USA and Canada
by Routledge
711 Third Avenue, New York, NY 10017

Routledge is an imprint of the Taylor & Francis Group, an informa business

British Library Cataloguing in Publication Data
A catalogue record for this book is available from the British Library

Library of Congress Cataloging in Publication Data
Coles, Tim.
    Student's guide to writing dissertations and theses in tourism studies and
    related disciplines/Tim Coles, David Timothy Duval, Gareth Shaw.
        p. cm.
    Includes bibliographical references and index.
    1. Tourism—Study and teaching (Graduate) 2. Dissertations, Academic—
    Authorship. I. Duval, David Timothy, 1970– II. Shaw, Gareth. III. Title.
    G155.7.C67 2013
    338.4′791—dc23   2012021944

ISBN: 978-0-415-46018-7 (hbk)
ISBN: 978-0-415-46019-4 (pbk)
ISBN: 978-0-203-07878-5 (ebk)

Typeset in Times New Roman and Franklin Gothic
by Florence Production Ltd, Stoodleigh, Devon, UK

CONTENTS

FIGURES

TABLES

BOXES

ACKNOWLEDGEMENTS

As may be expected, this book has been some time in the making. We are grateful to Andrew Mould (Publisher, Routledge) for first suggesting the idea several years ago. We would also like to thank Andrew and his colleagues, Faye Leerink and Michael Jones (Desk Editors, Routledge) for their encouragement and patience while we completed what at first seemed like a simple enough project! As we have learned, codifying knowledge produced in different continents and from three systems of higher education in as accessible a manner as possible has been less than straightforward.

The book is, of course, based mainly on our experiences of advising students at the Universities of Exeter in the UK and Otago in New Zealand. Many of the ideas and approaches presented here have been applied and refined in their work. To all our students, thank you. In particular, we would like to acknowledge the following for their input on issues which feature in this book: Haifa Abdullah, Supalak Akkarrangoon, Nikolas Boukas, Jamie Dallen, Lysianne den Breejen, Jessica Donald, Emily Fenclova, Rebekka Goodman, Feng-Yi Huang, Fiona Hutchison, Hanna Janta, Jian (Alexandra) Ju, Mojdeh Jamnia, Joo-hee Lee, Sotiroula Liasidou, Kristyna Marcevova, Jan Mosedale, Neil Ormerod, Pimlapas Pongsakornsilp, Fanny Raab, Catherine Salt, Hiroe Shimbakuro, Eleanor Stevens, Vibekke Stokke, Tea Tverin, Emma Wild, Julie Wooler, and Anne-Kathrin Zschiegner.

This book is about knowledge production and knowledge transfer, and we have been fortunate to benefit from insights on advising and assessing students from Adrian Bailey, Stewart Barr, Dick Butler, Claire Dinan, John Fletcher, Alan Fyall, Michael Hall, Adele Ladkin, Cassie Phoenix, Markus Pillmayer, Emma Roberts, Jarkko Saarinen, Nicolai Scherle, Karen Thompson, Dallen Timothy and Allan Williams. At the University of Exeter, Mandy Love and Heather Makin helped with the manuscript. The usual caveats apply.

The photographs are © Tim Coles.

1
STARTING OUT

Learning outcomes

By the end of this chapter you will know:

- What a dissertation is.
- Why doing a dissertation can benefit you.
- Why dissertations are important components in academic programmes.
- How this book works.

Introduction

You are some way through your studies. You may be dreaming about pulling out your best clothes ready for graduation, but there's one thing standing in your way. You have to produce a dissertation. Thousands of words. Independent research. A topic of your choice. Many weeks of work, perhaps spread out over an entire year. You may have done project work before, but this is unlike anything else you've ever faced. It is on an altogether different scale, many credits are at stake and, quite frankly, it appears a little scary.

These are pretty common and understandable emotions that many students encounter when they are first presented with an independent research, thesis or dissertation module. By the end of the process, these often – but sadly not always – give way to feelings of immense satisfaction and achievement, even outright relief, as the many hours you have invested in your research result in a great piece of work, a strong contribution to your degree result, and a boost to your self-esteem. More than a means to an obvious end, the dissertation can be an empowering and uplifting experience. You can demonstrate that you are able to conduct research on your own.

Aim of the book

Over time and with the help of this book, we hope you will find doing a dissertation enjoyable, perhaps even the highlight of your studies. Your dissertation needn't be a massive hurdle to overcome in the quest to graduate. The principal aim of this book is to provide you and other students with a suite of practical advice, guidance and suggestions for the preparation, execution, writing and completion of dissertations in tourism studies (and related disciplines).

What is a dissertation?

So far, we have taken the term 'dissertation' for granted. You may have heard other terms such as 'thesis', research 'project' or 'report' used. Your institution may even use some of these terms in your programme. In this book, we use 'dissertation' to denote a substantial piece of independent research led and directed by you which contributes significantly to your final degree result. This piece of work is likely to be undertaken in your final year of studies, but it may be started in your penultimate year. More than an extended essay, the dissertation ranges in length from 8,000 to 15,000 words depending on the institution and the number of credits it bears, perhaps being most typically in the region of 10,000–12,000 words at undergraduate level. In many programmes the dissertation is viewed as a pinnacle of a student's learning experience and the portfolio of taught modules is designed to enable or prepare students to conduct their own independent research. This is an important point to which we will return later.

Terms like 'dissertation', 'thesis' and 'report' are often used interchangeably. It may not surprise you to learn that there are contrasting and contested definitions. For instance, in one framework (Clewes 1996) 'project' is used to refer to independent research work at the undergraduate level (i.e. towards a BSc or BA), 'dissertation' is used for such work at master's level (i.e. MSc, MA), and 'thesis' for research-only, higher degrees (i.e. MPhil and PhD).

There are also important cultural issues and differences connected to terminology. This book is intended to be of use to students in countries whose higher education systems adopt Anglo-American models. There are, though, differences in usage between UK and North American programmes. In the case of the latter, the word 'dissertation' routinely connotes a text produced at the end of the PhD process. This is because many undergraduate and master's courses do not include independent research as a learning process. Students who do face this challenge at *undergraduate* level in North America – whatever the name of the exercise – will find the book of use.

So, too will those in European universities that have transformed their degree structures as a result of the 'Bologna Process'. The free exchange of people between member states is one of the central missions of the European Union (Coles and Hall 2005). The 'Bologna Process', which has been in operation since 1999, has set out to harmonize higher-education qualifications across Europe in order to enhance labour mobility (Munar 2007). Differences among qualifications in terms of their expectations of graduates were perceived as barriers to further closer integration. In Germany, for example, the term 'dissertation' was originally used to refer routinely to a doctoral thesis (i.e. for a PhD) and undergraduates would undertake a 'Diplomarbeit' (literally a piece of work for a degree). However, new qualifications and pathways have emerged from the so-called 'Bachelor's and Master's' system which seeks to replicate Anglo-American models of four-year study, with three years of a bachelor's degree followed by a fourth year to complete a master's degree. Independent research features strongly within the new system.

To be clear at the outset, this book is designed primarily to support undergraduate students in their first attempts to complete substantial pieces of independent research. The book is also specifically intended to make the dissertation more accessible to those studying travel and tourism in English as a foreign language (EFL). Higher education is a globalized sector, with ever greater numbers of students choosing to study abroad and/or in a second language.

This book attempts to describe many of the routine features of dissertation-based study which it is not always possible to discuss in depth in (shorter) institutional handbooks and briefings. It is important to note that this book is *not intended to be a substitute* for the specific guidance or detailed conventions you receive within your programme. You should diligently consult the handbooks and codes of practice that your institution publishes to govern your studies. You should be aware of – and more importantly, apply – your institution's policies and directives on such issues as plagiarism, referencing, ethics, health and safety. Of course, your advisor/s will be able to help in this regard, but it is your responsibility to be informed of such matters.

Those studying at doctoral and master's level will find a great deal of important and helpful information. After all, as we have noted previously, some readers' first encounter with independent research may be at master's level at the end of taught programmes. For those who have been through the process at least once, the book should serve as an important 'refresher' and it may even offer hints, tips and guidance that add to your stocks of knowledge and experience.

In fact, many of the central ideas and approaches introduced in this book are also directly applicable to more advanced studies. The central differences between an undergraduate dissertation and a master's or doctoral thesis are not routinely to be found in the tasks to be undertaken; rather, they are in the level of execution and in the degree of scrutiny. All the components that make a successful master's or doctoral thesis should be evident in their undergraduate equivalent (and vice versa). Expectations of master's and doctoral work are clearly much higher, the assessment criteria are usually different, and they are more stringently applied. For instance, independent research has to be located in the relevant academic literature (see Chapter 4), whether it is at undergraduate level or for a PhD. While undergraduates may be expected to critique the literature, they are offered a degree of latitude not available to doctoral students who are expected to demonstrate a fuller, deeper and altogether more critical appreciation, as befits a prospective expert in the field (which is an expected outcome of the doctoral process).

Why do a dissertation?

When we ask undergraduate students this question, the most frequent answer is – perhaps not surprisingly – 'because we have to!'.

This may reflect the compulsory nature of independent research in many tourism programmes. However, in others this may not be the case and the opportunity to undertake a dissertation may be optional. If you are contemplating whether or not to elect to do a dissertation, there are several reasons that make it a potentially wise choice:

- Dissertations are one of the few opportunities in your life to study a topic of your choice in depth, over an extended period of time.
- In many higher education institutions, dissertations are seen as a sign of 'graduateness' and a more representative indicator of the true 'exit trajectory' of students. As clumsy as these pieces of jargon may sound, they basically mean that the dissertation is the fairest and ultimate reflection of your all-round range of skills, knowledge and abilities right at the end of your programme.
- Successful dissertations may provide a significant boost to your degree result because they often account for a significant weighting of the final credits. Clearly, the dissertation has the potential to be a 'double-edged sword' because poor dissertations can have the opposite effect. However, the potentially positive contribution should focus your mind and galvanize your performance.
- Employability – you may be able to enhance your career or job prospects by undertaking detailed research on a topic, problem or field which is of interest and/or direct relevance to a prospective employer. At interviews, your dissertation may be a point for discussion and a platform for you to evidence both a range of pertinent skills and the ability to achieve consistently at a relatively high level of performance.

Approach of the book

In our experience, so many students – at least at the outset of the journey – believe that their dissertation may well be judged in its entirety without regard for how the various parts have been individually executed. This is the academic equivalent of saying that a chain is only as strong as its weakest link. Many a time we have heard students comment that this or that section is poor in their view so the marker must inevitably judge the whole dissertation as weak. Plainly put, this is why some students get stressed by the prospect of doing a dissertation and start to develop negative preconceptions of how difficult or, worse, seemingly impossible the task is.

We cannot say for certain that such a crude summary approach to assessing dissertation no longer exists. Check your regulations to be sure! What is more, we cannot legislate for those examiners whose ethos is that 'the glass is half empty rather than half full'. However, in our experience assessment has been increasingly moving away from fixating on and penalizing failures, limitations and shortcomings. Instead, the emphasis has been shifting more towards acknowledging (and hence appropriately rewarding) a student's skills, knowledge and expertise, and this philosophy should underpin how you approach your dissertation and it is a view we adopt in this book.

Articulated marking schemes have become commonplace, and they demand that assessors form a rounded view of the dissertation based on a series of criteria and/or questions (see Chapter 15). In some cases, the final mark is the sum of the marks earned for each task, sometimes with different tasks or sections offered greater or lesser weighting depending on their perceived relative importance. A highly consistent performance across the entire dissertation is most desirable, and the stronger this consistent performance across the piece, the higher the final mark is likely to be. Notwithstanding, it is possible to produce a successful dissertation in which some sections perform much stronger in relative terms than others.

Hence, you should not perceive the dissertation as an overbearing, massive task that is a case of 'all or nothing'. In this book we take the approach that it is best to deconstruct your dissertation into a series of smaller, more manageable tasks based on the standard components of a dissertation (see Chapters 2 and 13). You should concentrate on the operations necessary to successfully complete each task. Obviously, the components are connected and in some cases the tasks they entail do overlap.

By disassembling the dissertation into its constituent elements, you will find it easier to project manage as well as to make regular and strong progress towards its completion and submission by your deadline. The precise way in which you deconstruct your dissertation and sequence your work will ultimately depend on your working preferences as well as your institution's requirements. However, there are many common denominators that feature in dissertations across the world. Slightly different names or terms may be used but in Chapter 2 these components and their roles are introduced.

Put in reverse, as Chapter 13 indicates, to assemble your dissertation is a case of ensuring that you have each of these components complete and that they are bound together by a clear and robust set of aims and objectives that provide the common strands running through the components (i.e. chapters) and that are central to the tasks required to deliver each component. In one sense, then, the aim/s and objectives are the DNA that runs through the body of the dissertation. To extend the biophysical analogy further, certain parts of the dissertation act as vital organs: without them, or without them functioning properly, the dissertation will suffer dramatically. We will return to this idea in Chapter 2.

Features of the book

The book is designed to help you empower yourself to take greater control and responsibility for the successful completion of your dissertation. It utilizes several design features to help us debunk some of the popular and most pervasive myths about dissertations.

Proposal as a pivotal moment Cumulatively, this book is based on nearly 60 years of practical experience in helping our students to produce dissertations and theses at all levels of higher education. Much of this book is based on our observations of the particularities of putting principles into practice with our students. Many a time we have been struck by the disparity between how the principles of research are presented as compared to our students' experiences in reality. Research rarely runs as smoothly or problem-free as many accounts might suggest. Quite the opposite: it never ceases to amaze us how individual projects throw up new and interesting variations on common issues. It is, therefore, a key skill as an independent researcher to be able to predict, plan for, and respond to potential contingencies. Beyond a discussion of why each step is important to the overall goal, we attempt to offer you practical advice on how to plan, structure, write and review the components of a dissertation. We believe that, in order to do something well, it is crucial to have a clear idea of its importance and position within the wider scheme of work. In our view careful preparation is most vital to the ultimate success of your dissertation. For this reason, we emphasize the proposal as a key success factor in moving your research to what hopefully will be a successful conclusion. In our experience many students undervalue the work on their proposal and instead perceive the (usually longer) period post-proposal as most vital to their overall success, because this is (in their words) 'when they are really doing their dissertations'. While there is no doubt this too is an important phase that has to be conducted as competently as possible, a well-prepared proposal establishes a strong foundation for what is to follow.

Taking ownership and taking advice Throughout, we prefer to use the term 'advisor' not 'supervisor' with respect to the academic staff member/s (i.e. faculty) supporting you through the process (see Chapter 10). This is deliberate. We view the dissertation process as a journey of discovery, and our role – like that of your advisor/s – is to guide you along the right path. In our view the term 'advisor' is more commensurate with the concept of independent research in the social sciences, arts and humanities where tourism is studied. An advisor is someone who offers you support, guidance and advice which you can accept or reject along the way (with the associated risks such decisions entail). A dissertation is supposed to be *your* piece of independent research conducted by *you* and not the staff around you – that is, it is supposed to foster *your* abilities to generate a relevant research problem, to understand its context, and to design *your* schemes for data collection, analysis and reporting on the findings. In short, you should take ownership of the exercise. In contrast, the term 'supervisor' connotes someone who watches over a research project more closely and who provides input of a more formal, compulsory nature. This term is perhaps far better-suited to the physical sciences and laboratory-based work where a more structured approach is necessary because the dangers for inexperienced researchers are much greater and/or investment in equipment is much higher.

Research citizenship Your taking ownership of your project is one of the most important learning experiences of the entire dissertation process. However, with the rights to conduct a project of your design come certain responsibilities to your institution, your advisor/s and your participants among others. In this book we encourage you to consider both sides of this duality. In particular, you should be proactive in recognizing where you have responsibilities to others and how to discharge them most effectively. This is the reason why the relationship between you and your advisor/s has been left until relatively late in the book (Chapter 10). Many readers might think that this chapter should have come much earlier, and perhaps it could have done so. For instance, as we note in Chapter 3, previous research has revealed how tricky it can be for students to select a topic even with assistance. However, it is through the proposal that ownership of your project is asserted and we encourage you to be as independent as possible to that point.

Reducing your risks by breaking down the tasks There is no escaping the fact that many students think a dissertation is a risky undertaking because of the many credits that can be at stake. The advice provided here is intended to help you reduce your anxiety and lessen your risks. By breaking the dissertation down into its constituent components, you reduce the risk of forgetting to carry out a key task or leaving yourself too little time to conduct it properly. If you are encountering difficulties in conducting a task to the required standard, with the assistance of this book you should be able to identify the issues and obtain suggestions from your advisor/s at a much earlier point. Time is your most precious resource, but many students don't make the wisest use of it.

Case boxes and practical tips Throughout the book, we have provided case boxes based on experiences and issues that our students have encountered in their work. Most, if not all, of the problems or difficulties that you will come across are not unique to your research. These obstacles have probably been encountered in one form or another in the past, and they have been overcome. In many instances your friends and your peer-group are likely to be helpful for talking through matters and they fulfil a useful supplementary but different role to your advisor/s. By presenting issues faced by our students, we hope to provide you with a virtual peer-group as well as the reassurance that research does not have to be a long and solitary road to be navigated in the dark.

Learning outcomes and checklists There is a lot to remember when you are conducting a piece of independent research. At the beginning of each chapter, we set out the learning outcomes; that is, what you should know or be able to do after you have read the chapter and worked through the issues it has presented. At the end of each chapter, we summarize the main points and key issues to recap your learning with 'The chapter at a glance'. Alongside are checklists of things for you to consider and many of the common jobs for you to have done as a result. Of course, these are not exhaustive. Rather, they are intended to guide you in your progress in so far as the more questions you are able to confidently answer (and you can be the judge of that!), the better equipped and positioned you are to move on to the next chapter and stage of your research.

Structure of the book

Your dissertation is about process and procedure. It can be broken down into a series of tasks or operations and each one of them – whether examining the literature or conducting primary data collection – can be broken down further into stages. Thus, it is common to find the practice of conducting a research project presented as a linear process in flowcharts of varying degrees of complexity.

We also employ a flowchart as a structural device to enable you to monitor your progress towards completing the dissertation pathway (see Figure 1.1). More complex flowcharts than this no doubt do exist. However, our intention is *not* to attempt to map out the full intricacies of the many potential pathways, options and feedback loops that you *may* encounter. To be sure, research can be a much messier, more complex experience than this simple, straightforward device implies. You should be prepared for unexpected events, have appropriate contingencies in place, and we encourage you to think along these lines later in the book.

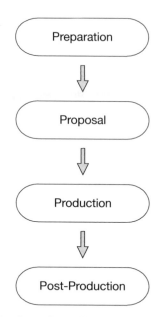

Figure 1.1 The four phases of undergraduate dissertations (and the structure of this book)

Source: authors

Instead, Figure 1.1 portrays four broad phases of work through which most under-graduate dissertations pass, namely the four P's of:

• **Part I: Preparation**

You need to familiarize yourself with what a dissertation actually is and the components that comprise it (Chapter 2). These are essential precursors to thinking through the tasks your dissertation will entail and devising a researchable topic by means of considering the background (Chapter 3), the existing literature (Chapter 4), and the research methods at your disposal (Chapter 5). Put another way, this part of the book deals with defining your research problem and setting the parameters for your study.

• **Part II: Proposal**

Before you start to work independently on your research, your higher education institution will scrutinize your ideas. Compiling an effective research proposal that communicates your intended research design clearly and in an acceptable format for your programme is an important step. Your research proposal may in fact feature in the assessment of your dissertation research. The components in research proposals are discussed in Chapter 6, followed by detail on how to plan your programme of work (Chapter 7) as well as discussion of your personal safety (Chapter 8) and the position of ethics in your research (Chapter 9). As important as time, safety and ethics are to students' research, they are often the subject of sketchy accounts in dissertation proposals. Time is an especially important resource which you should plan and use wisely.

- **Part III: Production**

After your research proposal has been approved you will be free to put your plan into action, but this raises several operational issues. Working with your advisor/s to obtain guidance and feedback is a crucial support mechanism and Chapter 10 examines how this relationship functions. Taking your ideas 'off the page' and putting them into action is the subject of Chapter 11. This expands upon ideas in Chapter 5 by examining some of the major practicalities for you to consider when collecting data. Finally in this part, Chapter 12 offers several suggestions on data analysis and interpretation as you start to write up your dissertation for submission (Chapter 13).

- **Part IV: Post-Production**

The production of a first draft is just the starting point of the final leg of your dissertation journey. To use a metaphor from the film industry, you have shot the film in its raw form; it is now time to edit and embellish it to enhance the experience. You are faced with a similar task in your work. Communication and dissemination are key skills for contemporary researchers. In particular, you are writing for an audience. As such, you should give careful thought to the precise appointment of your text (Chapter 14) as well as how your research will be examined (Chapter 15).

The chapters are deliberately designed to be short and readable. This is not a book we expect you will read from cover-to-cover in a single sitting. Rather, it is one you are more likely to dip-into from time-to-time as your work progresses. For this reason, we have provided a series of cross-references so you can easily find material in other parts of the book and so that you can follow particular aspects through the 'life course' of your dissertation. For example, we introduce aims and objectives in the next chapter, develop your understanding further in Chapter 3 on selecting a topic and drafting them in detail in your proposal (Chapter 6). Their role in designing your data-collection methods (Chapter 11) and analytical techniques (Chapter 12) are also covered later. Almost inevitably there is a little overlap as a result, but the major point is that the process of doing research is continuous, iterative and cyclical in nature.

What this book is not

Sadly, no textbook, however good it professes to be or how alluring the testimonials from famous academics on the dust jacket, can ever provide every single answer to every conceivable question on a particular topic. This is also the case here. We aim to be as systematic as possible in our coverage of the main steps, and in the process we attempt to be as comprehensive as possible in our advice about how to conduct the work towards your dissertation. If you follow the advice presented here, your chances of producing a successful dissertation should improve markedly.

Nevertheless, this book does not provide you with the perfect 'recipe for success', metaphorically-speaking. That is, if you follow every step outlined in this book there is no automatic guarantee that the perfect dissertation will be the outcome. Even if you buy the latest cookbook from a famous celebrity chef, there is no guarantee that your dish will turn out exactly or as good as it looks on page! You need to exercise discretion. The great majority of our advice will resonate with, or be applicable to, the dissertation pathway at your institution. However, you need to look at the instructions and guidance you are given locally and check how they relate to what is contained in this book.

Almost inevitably there will be some variations. For instance, as we will discuss later, there are considerable differences in practices and expectations, for instance in the way dissertations are structured (see Chapter 13) and examined (see Chapter 15). Sometimes the text alone is assessed while in other places proposals, texts and even verbal presentations contribute towards the final mark. For some institutions secondary data is sufficient to inform the dissertation. In contrast, many programmes demand primary data collection and analysis so that a student can demonstrate the wider range of skills and abilities needed to design and conduct independent research. Indeed, this is the premise we have taken in this book.

Finally, this is not a book about research methods and techniques; it is a book about the dissertation *process*. Quite often books on methods are couched in the process of doing research because the two are, understandably, connected: methods are an important means to an end (completing a research project), but they are not the sole means.

There are many detailed textbooks and reference guides to quantitative, qualitative and mixed methods research, some specifically intended for tourism students (Finn et al. 2000; Phillimore and Goodson 2004; Ritchie et al 2005; Brotherton 2008; Buglear 2010; Baggio and Klobas 2011; Hall 2011a; Veal 2011) which you should consult if you are looking to learn more about methods in depth or to refresh your prior learning.

Indeed, this latter point is very important. In practically every other module you have studied the design of the curriculum and assessment will have been linked. This is usually set out in so-called 'intended learning outcomes' (ILOs). You should have been assessed only on skills and knowledge for which you have received direct training and/or your learning has been actively fostered. In other words, you should not be assessed in areas where this has not happened because this is a somewhat unfair test. The dissertation should be no different and it should have ILOs. Before expecting you to conduct independent research – especially that requiring primary data collection and analysis – you should have received adequate training on research methods and techniques as part of your programme. You may not necessarily have received this as part of, say, a dedicated series of briefings for the dissertation module. Rather, it is more likely to have been either in dedicated modules or embedded within other thematic modules. Wherever or whatever the mode of delivery, you should revise your prior learning as an essentially preparatory exercise. If you believe that you have not been trained in research methods and techniques to a level adequate to support your dissertation, you should discuss your concerns with the programme staff before starting your work. You may want to reflect upon why you are doing a dissertation, especially if it is optional.

This book is more about the 'why' and the 'how' of conducting a piece of independent research in tourism, and about how, by linking the various components and stages in the process, the whole will be greater than the sum of the parts. The guidance we provide about methods (Chapters 5, 11 and 12) is about encouraging you to make the appropriate choice of methods to address the aims and objectives you set yourself after selecting your topic (see Chapter 3) and reviewing the extant literature (Chapter 4).

The chapter at a glance

The main learning points of this chapter are that:

- Your dissertation is intended to reveal the breadth and depth of your knowledge of and expertise in independent research.
- You have to take ownership of your dissertation and you alone are responsible for completing the dissertation journey.
- There is no single perfect way of doing a dissertation.
- The best way to approach the work for your dissertation is to break it down into a series of more manageable tasks.

Dissertation checklist

Before you go further in your work, check you:

1.	Understand what a dissertation is and how it helps you develop academically within your programme of studies.	
2.	Have a copy of the 'dissertation handbook' for your programme.	
3.	Know what you are expected to have achieved by the end of the dissertation process.	
4.	Attend all preparatory briefings.	
5.	Know how long it is going to be (words), how long it should take (time), and what proportion of your final mark it comprises.	

Part I
PREPARATION

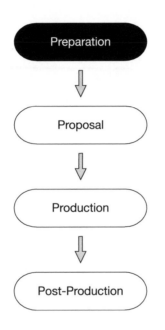

2
SOME BASICS ABOUT DISSERTATIONS

Learning outcomes

By the end of this chapter you will be able to:

- Identify the components of most dissertations.
- Explain how the components of the dissertation are linked.
- Recognize the importance of strands of continuity running through your work.

Deconstructing your dissertation

In the previous chapter we started to look at the various challenges that a dissertation presents you with. A dissertation is a significant piece of self-directed work in which you will demonstrate your skills, knowledge and competence in conducting independent research. Although it may appear to be a dauntingly large exercise to attempt, we have suggested that the best approach is to break the dissertation down into a number of more manageable tasks related to the major components of the dissertation. Each of these tasks can be project-managed on its own. By adopting this sort of approach, not only will you be able to make more regular progress, but also you will learn to monitor and evaluate your own work more effectively and continuously. You may not realize this, but monitoring and self-evaluation are vital skills to take from the dissertation to the next step in your career path.

There are two aims for this chapter. The first is to disassemble the dissertation into its constituent elements and in the process to explain the function of each part. If you understand more fully the roles played by each part you are, first of all, better placed to design your work (see Chapters 3–5), to set realistic, actionable aims and objectives (Chapter 6), and to establish a feasible project management timetable for your research (Chapter 7). Not all parts of the dissertation require equal time or attention to be devoted to them. This is another way of saying that there are some parts of the dissertation that need to be prioritized over others. This is because they are likely to feature more heavily or prominently (notice the use of the relative) in the assessment of your work (see Chapter 15).

The second purpose is to identify how the key components in the dissertation link together, and how the linkages offer a means of establishing strands of continuity and connectivity to ensure that a much stronger piece of work results. We propose a basic triangular model of how the main parts of the dissertation function together.

By way of a clarification, some programmes may require their students to divide their final texts into 'sections' rather than 'chapters'. By and large, this is just a question of semantics and preferences. In some cases, though, it may reflect the applied traditions of, or vocational orientation to, many tourism programmes. Consultancy or contract research reports for businesses and organizations are commonly divided into 'sections' rather than 'chapters'. Here we use the term 'chapter' in keeping with a more purely academic approach (and 'sections' and 'sub-sections' for the subsequent division of a chapter).

Principal components of a dissertation

The following comprise chapters in dissertations around the world, although they are not always present in every dissertation and nor is this particular sequence always followed:

- Introduction
- Literature Review
- Background
- Methods
- Results
- Discussion
- Conclusion

The introduction chapter sets the scene and it outlines the basic parameters for the dissertation. Key ideas are proposed, central concepts are defined, and some initial background information is provided. In the case of the latter, this may be remarks on where, when and why the research was undertaken. The introduction serves three main purposes. First, to set out the rationale for the research. In effect, this is the justification – that is, the reasons why the work is important or timely, as well as the contribution it will make to studies of tourism. Second, perhaps its most crucial function is to articulate the aim/s and objectives – as key threads of continuity – for the first time. Third, it alerts the reader to the structure of the text by providing a kind of 'roadmap' through each of the chapters.

The purpose of the literature-review chapter is to examine extant work on a topic area. It sets out to establish the *intellectual* case for the dissertation more fully (i.e. beyond the preview in the introduction) and it serves to justify the choice of aim/s and objectives more specifically. Previous contributions may have been published in tourism studies as well as other disciplines, most notably within the social sciences (cf. Holden 2005; Coles et al 2006). Hence the literature review both informs and positions the work, and it helps to establish the originality and likely contribution of a dissertation.

The review may take the form of identifying theoretical, conceptual or thematic ideas or trends. Often overlooked, though, is the fact that academic literature can be examined for precedents in terms of the methods and strategies for data collection and data analysis as well as for the type of locations in which a topic has previously been studied (see Chapters 5, 11 and 12). Findings from the extant body of knowledge should contextualize the results and drive the discussion of findings in every dissertation.

Readers have to be informed of key contexts: the 'real-world' rationale, perhaps in terms of trends and events in policy, economy, society, culture and the environment, or a combination of them (see Box 2.1). This is distinct from the intellectual justification rooted in the academic body of knowledge and presented in the literature-review chapter. Information from reports, statistical digests, newspaper articles and the like establishes the importance of a topic beyond academia and further cements the case for studying it. Recent developments in policy and/or practice frame many of the issues studied in tourism, and one of the enduring critiques of tourism studies has been the highly applied nature of much of the research in the subject area (Franklin and Crang 2001; Coles 2004).

Whatever the merits of such arguments, this background also acts as further context for the results and conclusions later in the dissertation. It is perhaps more usual to find the background discussed in the introduction in undergraduate dissertations. However, the degree of contextualization may necessitate a separate chapter – for instance, where there is a protracted history or the issues are especially complex. Background chapters tend to be used more in master's or doctoral work but undergraduates can use them if needed or if regulations and word count allow.

The next chapter which features in every dissertation sets out the methods and techniques used to collect and analyze data in order to address the aim/s and objectives. This chapter is a record and a defence of the decisions made in designing the research and its analysis, as well as a prelude to later discussion of the results and findings. To avoid confusion, although it is necessary contextualization for the results chapter, discussion of the methods does not form part of the background.

Many students routinely refer to this chapter as the 'methodology', not 'methods' or 'methods chapter'. The distinction is important. Methodology refers to a branch of the philosophy of research connected to epistemology (i.e. in plain English, the nature, scope and production of knowledge) and ontology (i.e. simply put, how what exists is understood, viewed or made sense of) (see Tribe 2009). Quite importantly, as Del Casino et al (2000) demonstrate with their multiple readings of a visitor attraction in Kentucky, particular methodological choices and approaches – which are driven by allied epistemo-logical and ontological decisions – result in differences in the collection, analysis and interpretation of data on the (same) particular phenomenon. Very often undergraduate students are not required to, or chose not to engage with, these higher-level theoretical concerns. Instead, they provide a description of, and justification for, the type of approach (i.e. quantitative, qualitative; exclusive, mixed or multiple methods) and the particular methods and techniques they adopt.

Thus, the methods chapter heralds the start of the second 'half' of the dissertation and it is usually followed by a results chapter or chapters (i.e. plural). Whether or not a single results chapter is employed or there are several chapters containing results will depend on many factors, including: the nature and volume of the data collected; the type and

Box 2.1 Background, literature review and rationale: in search of connections

The last decade has seen the emergence of leadership as a distinctive aspect of, even sub-discipline, within management studies. In broad terms leadership is considered distinct from management in so far as managers have responsibility for administration and the functioning of an organization, whereas leaders set its vision, mission, strategy and ultimate direction. So, the thinking goes, leadership drives the success of an organization and, moreover, it defines what success looks like (i.e. profit maximization, potential community engagement, minimizing environmental impact).

Tourism organisations have not featured in leadership studies and what's more this subtle but important difference has for the most part been overlooked by the tourism academy. The latter is perhaps all the more surprising because many academics working in the broad area of sustainable tourism have exhorted consumers and those working in the 'industry' to show more leadership by acting with greater responsibility as a means of stimulating more widespread behaviour change.

In addition to this academic rationale for studying leadership in tourism organisations, one of our students discovered that there were strong reasons rooted in practice that further justified a detailed study on the relationship between different styles of leadership and sustainable business in the South West of England. As one of the most iconic tourism attractions in the United Kingdom, the success of the Eden Project in Cornwall has been commonly attributed to the values, vision and determination of Tim Smit. A closer reading of his autobiography demonstrated that he adopted a largely transformational rather than transactional style of leadership; that is, change is primarily motivated by values rather than instrumental considerations.

Other popular media sources such as interviews and features in newspapers as well as various web pages triangulated this view and furthermore pointed to the role of leadership in the approaches of the Bedruthan Steps and Scarlett hotels, also in the county and identified as sites of innovation and best practice in sustainable tourism management. Thus, in addition to the published academic literature for the review chapter, a careful survey of the popular media and grey literature revealed plenty of other sources that could be used as supporting evidence for a background section in the introduction.

Source: authors

level of analysis; and the regulations governing the format of dissertations. The results chapter traditionally presents the outputs of data analysis as well as interpretations of the data in the form of findings.

Traditionally, the conclusion provides a summary of the main findings from the research as well as the limitations of the study and its implications for further research on the topic or in the field. The conclusion is an opportunity to reflect on the research process but students traditionally struggle to identify the limitations and implications of their research (Chapter 13).

Finally, you may have spotted that the discussion chapter has been missing so far. Don't worry: this is deliberate! As we note later (Chapters 12), 'results' are not exactly the same as 'findings'; in short, findings are results in context. Put another way, data are only meaningful when they are set in context, whether that is established by comparison with material cited in the literature review (i.e. academic) and/or background section/chapter (i.e. practice and policy).

Separate discussion chapters offer the space and scope to address these issues independently and away from the main body of data. As such, they tend to be used more in highly quantitative studies (Chapter 12). Still, for all their potential to add value, and just like background chapters, discussion chapters frequently do not appear in undergraduate dissertations mainly for reasons of restricted word count and/or because they are not encouraged by the regulations. Instead, the discussion of the findings is absorbed into the results chapter/s and the conclusion. Once again, the purpose of mentioning this type of chapter now is to raise your basic awareness of it. And for reasons of consistency with common practice at undergraduate level, we adopt this much simpler position in Chapter 13 when we discuss writing up.

Dissertation from A–Z?

Many students have the preconception that, just like correctly reciting the alphabet, the chapters of, or work towards, the dissertation must be completed in a simple linear sequence, from the introduction to the conclusion culminating in the final reference list and any appendices. Of course, there is nothing stopping you from working in this manner if you want to. It is certainly very methodical and could make for a straightforward approach to managing your project.

In fact, it is almost always the case that this obvious sequence is not followed: instead, you have to show a level of flexibility in tasking your work and you may need to undertake several tasks at once in parallel. Indeed, there are some sound reasons why you should deviate from this path. Most relate to the complex demands and basic realities associated with completing each of the components in your dissertation. For example, you may wish to write or at least complete your introduction last of all. If you think about it, how can you complete an explanation of what is going to be in each of the other chapters until you have completed them?

Some students start their writing with the literature review, not least because they have been reading extensively from an early point in the preparation of their proposal (see Chapter 6) and they think it will take longest. However, many others start the process of writing up with the methods and then results (data analysis). This is because they perceive the literature review as a greater challenge to their writing skills, they see the methods and results chapters as connected and more straightforward to compile, and they want to develop some confidence by getting into the 'flow' of writing before they attempt trickier tasks. Others compile the methods and results chapters first because the material and key decisions are fresh in their memory. What is more, they wish to craft the rest of the text around them. In some cases, though, this is not for strategic reason of ensuring continuity. Rather, for purely instrumental reasons they want to make sure they have enough space for these two chapters and allocate the as yet 'unused' words to the remaining chapters.

Whatever the sequence, these precede discussion of the background – either as a section of the introduction or perhaps even as a separate chapter of its own – and the main findings (i.e. part of the conclusion). It is very rare these days, especially in tourism studies, to be able to take pioneering footsteps marking out the boundaries of new intellectual terrains, because there is so much new knowledge being produced so rapidly by a burgeoning academic community. As such, you will most likely have to contextualize your results based on existing academic precedents and analogue studies or progress in policy or practice. In turn, this may suggest to you further avenues for future research that should be usefully discussed in the conclusion which, like the introduction, is one of the last components to be completed.

Once these chapters are complete, you may though need to revisit some of your earlier drafts depending on your timeline for writing up. Dissertation research takes place over several months, during which time new contributions may have been published in the academic and practitioner communities which may further inform your literature review or background.

Thus, before we move on further, there are several observations that are worth highlighting at this point:

- To complete your dissertation successfully, it is best to think of it as a series of components (i.e. chapters or sections).
- Each chapter in your dissertation will, in turn, require you to complete a number of tasks.
- You will have to demonstrate a range of both subject-specific and transferable knowledge and skills in order to complete these tasks.
- Within certain limits, you can be flexible in terms of the sequence in which you conduct these tasks.
- A more flexible approach will, however, require you to think through, plan and project-manage your research more carefully.
- There may be variations in the sequence of chapters and/or where some material is placed in your work but, by and large, each of the main elements outlined above should feature in a successful dissertation.
- The more of these tasks that you successfully complete and to the highest standard, the greater your final overall grading is likely to be.

Linking the components together

So far, there has been one major implicit assumption in the discussion of the main components of the dissertation: namely, that all of the chapters will be of equal importance to you and hence the readers (i.e. assessors or examiners). We have also implied that the chapters will be linked. Some of these overlaps have been identified briefly above and these connections will be discussed in more depth as we progress through the book. Cross-references from chapter-to-chapter, and section-to-section, are one obvious means by which you can engineer a degree of continuity. Sometimes material discussed in depth in one chapter may necessarily have to be repeated or précised elsewhere in one or more further chapters (as indeed is the case in this book).

Nevertheless, making connections should not be a casual or even serendipitous affair. Rather, we would argue that you should seek to weave very deliberate and robust strands of continuity, logic and argumentation through your dissertation (see also Chapter 13). As Figure 2.1 illustrates, these strands relate to the mutually-reinforcing connections between the literature review, the methods chapter and your data analysis (i.e. results).

These linkages are facilitated through the aim/s and objectives you first articulate in the introduction as a preview to the rest of the study. This relationship starts to function when you have some ideas of what you would like to study (Chapter 3):

- The literature review helps you to locate and contextualize your topic in the existing thematic body of knowledge on the subject. Your reading may be in tourism studies and/or in other disciplines (mainly in the social sciences). Your review of the current state-of-the-art on your subject helps you to refine as well as justify your aim/s and objectives.
- Your choice of methods has to be entirely appropriate to your aim/s and objectives, otherwise at best you will generate spurious data or at worse results and findings of

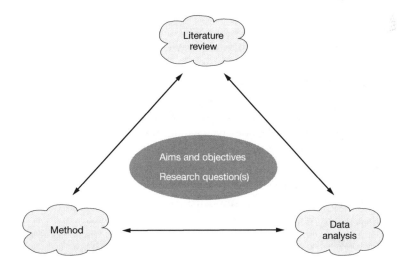

Figure 2.1 Connections within your dissertation

Source: authors

quite dubious relevance. There are important linkages in so far as your precise choice of approach and instruments will also be informed by precedents in previous studies which the literature review should reveal. These may even necessitate a review or redrafting of your aim/s and objectives. Beyond the thematic literature, further review of the latest specialized methodological texts will inform your selection of methods and techniques for data collection and analysis.

•   Your choice of analytical techniques, processes and measures has to be appropriate to the types of data you collect and the methods you have used to generate the data. They must allow you to address your aim/s and objectives directly. There is no point in generating unfocused results. Backward linkages are important in the sense that the comparison of your data with those contained within previously published studies cited in the literature review contextualize your results and establish the relevance, significance and contribution of your findings.

Thus, after they are set out in the introduction, the aim/s and objectives should serve to choreograph the remaining chapters, and they should be central to the strategic and tactical decisions you make in conducting each task within your research.

Less successful dissertations are often characterized by serious dislocations between the aim/s and objectives in the introduction as compared to the results that are reported and/or the conclusions that are drawn from a study (see Chapters 13). If your conclusion does not directly address the aim/s and objectives from the introduction, you may have a dissertation of two distinctive and very different 'halves'.

The dissertation is like . . . a good novel?

It is probably fair to say that for most people the words 'dissertation' and 'novel' may not immediately be associated. Dissertations are not always renowned as good reads. Then again, that is not necessarily their purpose. However, while good style will certainly do you no harm and may even be welcomed by your examiner/s and other readers, the connection is not as far fetched as you may imagine.

Figure 2.1 is at the heart of how we advise our dissertation students and it is central to how we want to help you in the remainder of this book; that is to say, the advice presented in Parts I to IV is intended to enable you to establish, maintain and benefit from this relationship in the completion of your dissertation. It is hardly surprising, then, that a clear, compelling introduction and a strong, confident conclusion are vital chapters of your dissertation. The other pivotal component in your dissertation is your methods chapter. This acts as a bridgehead from setting out your research and the case for your study, through the generation of relevant data, to the results, findings and conclusions.

Some students may find this surprising. There is no doubting that the literature review and results chapters are important, and they are often heavily valorized in dissertation assessment schemes, or rubrics as they are termed in North America (Chapter 15). Students often perceive the literature review as the chance to show off bibliographical and critical writing skills while the results chapter is the chance to demonstrate their analytical wizardry. However, our position refers to the process of producing a successful text overall. If the two exercises are disconnected then their effectiveness individually, and their contribution to the entire document, is constrained as a consequence. It is in this regard that the methods chapter occupies a pivotal position because it adds great value by demonstrating your ability to make appropriate decisions to link the extant body of knowledge with your current research.

So, the notion of the three-part novel has almost become a cliché, but the general principles hold true for dissertations. In the words of the well-known aphorism: tell them (i.e. your readers) what you are going to tell them (introduction), tell them what you are telling them (literature review–methods–results), and tell them what you told them (conclusion). This is a point we will return to in Chapter 13 when we discuss writing up.

The chapter at a glance

The main learning points of this chapter are that:

- **A dissertation comprises a number of interconnected components (i.e. chapters), each of which fulfils a distinctive role within your dissertation.**
- **Each component makes different and distinctive demands of you as an independent researcher.**
- **The aim/s and objectives represent important strands of logic, continuity and argumentation that should run through the entire body of your dissertation.**

Dissertation checklist

Before you go further in your work, check you:

1.	Know precisely which combination of components is expected in a dissertation within your programme of study.	
2.	Understand the role played by each component in your dissertation.	
3.	Are able to recognize the most important components within your dissertation and explain why they are significant.	
4.	Appreciate the difference between methods and methodology.	

3

SELECTING A TOPIC

Learning outcomes

By the end of this chapter you will be able to:

- Generate ideas for a dissertation topic based on reading and prior learning.
- Apply several criteria to assess the relative merits of your ideas.
- Select an idea to refine further into a research proposal.

Introduction: getting started

The choice of a viable research topic is of critical importance and one of the most significant stages in the dissertation process. In his early work on researching tourist satisfaction, Ryan (1995) pointed out that the first of eight stages in the research process is the identification of the research problem. Yet this, as well as being one of the earliest tasks in the dissertation process, is also one of the trickiest. As I'Anson and Smith (2004: 23) have observed, in part from personal experience, 'even with research methods teaching and support. . . . it can be overwhelming for the students'.

This chapter provides a framework to help you through this critical stage. In doing so it explores the criteria involved in the choice of a topic from three perspectives. The first concerns the assessment criteria used in the dissertation process from a formal, academic standpoint. In contrast, the second draws on a series of key criteria that involve your skill set and interests as a student. Of course, it is important that these two perspectives fit together. Finally, we recognize that many students without English as a first language conduct dissertation research in the English-speaking world. Where possible we attempt to highlight those areas of the selection process that may cause particular difficulties for international students. This draws on a range of studies that have drawn attention to specific 'cultural' problems encountered by some international students (Cadman 1997; Barron and Arcodia 2002; Lowes et al 2004).

Interests, motivations and expectations

Topic selection is tricky for a variety of reasons, not least of which is the nature of your training before you commenced work on your dissertation. Prior to this point, your training was largely dominated by lectures (Saunders et al. 2000). Tourism studies is a wide-ranging subject area and it crosses a number of traditional disciplinary boundaries within the social sciences. As a result, you can often be confronted by a bewildering array of potential topics for your research.

The usual advice to students at undergraduate (and master's) level is to choose a topic that you are interested in. You may have encountered a topic within your lecture programme, in seminar discussions, as part of your reading or perhaps from a presentation by a guest speaker which has fired your imagination. In part your choice may also relate to particular motivating factors such as personal connections with a particular study area or looking ahead to a possible career. Beyond the development of specialist knowledge, dissertations can often be used at interviews to evidence a student's interests and capacity for independent work (Webster et al 2000).

In a study of Asian students three key motivations affecting the choice of dissertation topics were, in rank order: links to particular career development; personal interest in a particular subject area; and perceived ease of access to data (Huang 2007). In wider research Cadman (1997) and I'Anson and Smith (2004) demonstrated the importance of personal interest. Cost factors and local knowledge are also important and legitimate considerations in light of the effort needed to complete your research – for many students, living and working back in their local area or country brings tangible advantages. Conversely, for others the novelty of working somewhere very different offers fresh challenges that stimulate their research interests.

While such motives are clearly of significance and are stressed by many advisors, they should be put into an academic context that draws other factors into play. There are several strong pedagogic reasons for conducting research projects (Collis and Hussey 2003), important among which are:

- Active learning through the identification of a particular problem to be investigated (i.e. the primary concern of this chapter), through to the completion of the work.
- Application of your academic knowledge (Chapters 4–5), and analytical problems-solving skills (Chapters 11–12) to address problems, generate solutions or develop a greater understanding of key issues.
- Further development of your (transferable) skills for future independent (research) work.

Matching your interests and skills with access to information

Clearly dissertation research is as much about your intellectual development as it is about your skills development (I'Anson and Smith 2004). Thus, developing an idea that covers both your interests and the pedagogical expectations of the dissertation is an important consideration. However, two further sets of issues need to be taken into consideration in your decision-making process: the skills that you have, and your likely access to data.

The first is, to a degree, not insurmountable and it should not always act as a barrier to your choosing a topic of personal preference. After all, you can acquire new skills or work to further develop your existing skills to allow you to tackle a particular research topic.

For example, you may want to undertake a relatively large survey of people's past involvement in the early stages of package tourism during the 1960s. While you may have done some semi-structured interviewing in the past, your preferred method is to use the richness of in-depth interviews through the oral history approach. Through this, you would discuss with people their past holiday habits and their recollections of their first package holiday. Not only might this require you to read up on a new, largely unfamiliar approach to data collection, but also it might require you to learn new skills for the data handling and analysis of the large amounts of transcribed information you generate. One of the most frequently used packages for this purpose is NVIVO (Bazeley 2007); however, as a dedicated piece of software for qualitative analysis, you will need to read specialist guides on it and attend some training courses in order to make the most of using it for your research. New skills development of this type should be factored into your timetable (Chapter 7), and justified in your research proposal (Chapter 6).

Clearly, the converse is also possible: if you are unwilling to work on overcoming your own current limitations, it is illogical to select a topic in which you are starting from a point of disadvantage. For instance, there is little point in selecting a topic that requires a high degree of face-to-face data collection with members of the public if you are very uneasy about conducting interviews or on-street questionnaire surveys.

Nevertheless, access to data is often the more difficult issue and it is more likely to be the insurmountable barrier. This is especially the case if you are considering, or are compelled by your regulations to conduct, primary data collection. Thus, even at the very early stages in your thinking it is important to identify potential problems of access that could make a topic unworkable. Many students have ambitious ideas that are only feasible if the data can be accessed. In no small part this depends on your skills, preparation time, and the willingness of others to cooperate with your work. What is more, you need to develop a realistic view of whether you will be able to get access to the data.

One obvious point is to remind you that you are not the only student conducting research and trying to get access. Yours will be just one of many programmes in tourism studies requiring its students to conduct dissertations. In addition to undergraduate students, there will be master's and doctoral students undertaking their research as well as a range of projects underway from school students and students in further and tertiary education. As Box 3.1 indicates, you have to resist the temptation to be insular in your thinking and you should consider the issue of access from the perspectives of your subjects or potential host organizations.

For instance, popular and high-profile attractions may appear to be appropriate subjects for, or locations to conduct, your research. However, in our experience students often underestimate just how many of their extended, external peer-group have had precisely the same idea! In many conversations, we have been struck by just how overwhelmed attraction managers have been by the sheer volume of requests they receive annually. Frequently, they have observed that many students have unrealistic expectations; they have not considered how their work might impact on the operation of the business and/or their visitors' experiences; and no benefits are presented in return for the cost and

Box 3.1 What price access?

It shouldn't come as much of a surprise that many students identify airports as a place where they want to conduct their questionnaires or interviews. Hundreds if not thousands of passengers pass through each day, many with time to spare after they have checked in, passed security, or entered the departure lounges to wait for their flights. What better for this captive audience than to break the tedium by spending a few minutes completing a questionnaire or being interviewed? After all, they have nothing better to do. So, it often comes as a bit of a surprise when we advise our students that it can sometimes be tough to get access to conduct their research at airports.

Airside access is difficult if not time consuming to arrange these days because of heightened security concerns. Airlines are working on increasingly tight turnaround times so they are less keen for research to be conducted in lounges or in departure halls. Concession holders prefer passengers to use their time browsing their stores and buying their wares. And in one extreme instance, an airport manager even reported that his team had been inundated with requests from students, many letters were poorly written and vague, and there was often no idea of precisely what was proposed. As a result they now had a blanket policy of no student research on the premises.

More reflective students take this advice on board. Some even decide to pursue slightly different topics while others develop contingency plans in the event that access is not granted. Sadly, some do not. In one particular case, a student was so adamant that access would be granted that there was no Plan B. After a polite and informative letter to the airport manager, the student had been invited to a meeting to discuss the work. Verbal consent had been granted but there the problems started. To get airside access, the manager had to write to a number of stakeholders, principal among whom were the border-control authorities and the airlines operating in the zones where the work would be conducted. This took time. Days passed and days turned into weeks and weeks turned into a couple of months. Despite the advisors' warnings to develop an alternative plan, this left just three weeks to collect, process and analyse the data as well as to write up. The dissertation was handed in on-time – somehow. Sadly, the quality suffered as a result. The data analysis was very sketchy, the methods section was a little hazy to say the least, and the markers agreed on a much lower grade than the student had expected. Access had been granted but the victory came at a heavy price.

Source: authors

potential reputational risk of their being involved. In short, students want the right to conduct research at given locations without understanding what their responsibilities entail (see Chapters 8 and 9).

Difficulties frequently emerge when researching businesses. Many tourism dissertations – and much tourism research in general – is focused on the tourist as the unit of analysis because tourism businesses can be quite unforthcoming with their time or in their answers to questions, perhaps for reasons of commercial confidentiality. A survey of 42 tourism students undertaking projects found that some 'students felt they had access problems because student research was not highly rated by organisations or individuals' (I'Anson and Smith 2004: 28). Huang (2007: 36) confirmed such views for international students and concluded that during 'the early stages of topic selection, the challenges of assessing and collecting data should be emphasised'.

Problems of data availability compound basic issues of access. While there has been a proliferation of research into small- and medium-sized tourism enterprises (SMTEs, Thomas et al 2011), there has been comparatively less research on larger-scale international businesses (Hall and Coles 2008). This may make such businesses appear attractive for dissertation research. As several contributions make clear, a topic that depends on obtaining information from senior decision-makers as well as access to such individuals can prove especially challenging (Mosedale 2007; Bochaton and Lefebvre 2011), perhaps even impossible for most undergraduates.

The nuts and bolts of topic selection

As this initial discussion demonstrates, you will need to adopt a more structured and systematic approach to topic selection, comprising three key stages:

1. Initiating the idea/s.
2. Linking your idea/s to other academic studies.
3. Thinking about the practical implications – is the study feasible?

Stage 1: initiating the idea/s

In the first stage, as our previous discussion has shown, this is motivated by your interests. These may have been stimulated by your prior studies, an internship or perhaps your travel experiences.

If at this stage you are still in need of inspiration then you should start by making a list of all those topics that you have found engaging during your taught modules. This should also take into account the material you have found especially interesting to read. Is there any pattern emerging from your list? For instance, are the majority of your topics associated with one particular theme?

Box 3.2 illustrates this process using possible areas of interest in the broad theme of tourist behaviour. As you can see, these have been given a little further detail. From initial concepts, the next step is to work up in a little more detail two or three potential topics. Remember, at this time you are not yet restricted to a single concept for your study and

---

Box 3.2 Sketching possible dissertation topics on tourist behaviour

---

- Young British tourists and their holiday behaviour.
  This could look at whether there are patterns of excessive, anti-social behaviour among young holidaymakers in British seaside resorts as media reports suggest.
- Motivations of backpacker tourists.
  This could be developed around the motivation and decision-making processes of potential backpackers in the UK, not from the UK.
- Group decision making and holiday behaviour among young people.
  Most tourist behaviour studies focus on the individual but many young people travel in groups, so this dissertation might look at behaviour in this different unit of analysis.
- Overseas students as tourists.
  We know many students study overseas these days but less is known about where they spend time when they are not studying and how they weave their trips into their study programmes.
- People with disabilities and access to holidays.
  This could focus on access in terms of pre-trip behaviour along with barriers to access.

Source: authors

---

instead you can develop multiple ideas if you are spoiled for choice or have no firm views on your intended direction.

In order to work up your idea/s, you could do something as simple as writing down in a very short time period (say 10–15 minutes) what you know about your potential topic/s and how they might be researched (see Box 3.2). This may also include an initial indication of where the topics may be researched; that is, the geographical location/s. In most instances dissertations are case-study based in so far as they are undertaken in a particular location due to constraints of costs, resources and time.

Another popular method is the use of mind maps. These resemble a tree, with multiple branches emanating from several large trunks: the former relate to specific directions for investigation while the latter represent the key issues or areas in your research. Mind maps allow you to visualize the extent of a topic and where/how one particular issue relates to another. Given that the extent of a dissertation can often be quite narrow, a mind map helps you to focus, to ensure that the scope of your work does not become too broad, or that your project is not too intricate.

Alternatively, you could try to identify a topic starting with its title. Of course, this will (or should) only be a draft or working title that will no doubt change and be refined as your work unfolds. In fact we would suggest that you try writing out a simple, draft title at this point in Stage 1. You could even write this as a simple question.

If you have trouble drafting a title, have a look at a variety of articles in some of the academic journals to see how other researchers have formulated theirs. Remember, do not make your titles too long or too complex; equally, do not make them too general. For example, going back to our ideas on potential topics related to tourist behaviour (see Box 3.2), a title such as 'How do tourists behave on holiday?' is far too general. We would need to think about what type of tourists, what types of holiday and of course what destinations.

Dissertation processes and associated support systems vary among programmes. However, staff (i.e. faculty) members – or the potential advisors of the future (Chapter 10) – may be willing to discuss ideas with you in the very early stages of your project development. That is, they may be permitted to discuss initial, provisional ideas before you undertake detailed work on your research proposal (see Chapter 6), especially if this is summatively assessed (Chapter 15). But remember, these discussions will be more fruitful if you approach them with more than very vague notions or a totally blank sheet of paper!

Stage 2: linking your idea/s to other academic studies

Stage 1 may culminate with the identification of a potential topic, or perhaps two or three possible options. Stage 2 involves refining your initial idea/s further within the more specific context of the literature (see also Chapter 4). At this point, your main concerns should be:

- *Whether this type of study has been undertaken before?* – Many students search for unique ideas. This is more the case at PhD level than master's and undergraduate level. However, given the large numbers of students studying tourism at all levels globally it is becoming increasingly difficult to find something that has not been examined already. More likely you may find a topic that has been relatively under-researched. Uniqueness is therefore not necessarily a criterion for the selection of a dissertation topic. If you find a unique topic then great, but if not you can still obtain high grades and do well.

- *To identify gaps within the literature on a particular research theme* – It may well be the case that closer inspection of the key debates in the literature of your selected research topic reveals small but significant gaps that may warrant further research. Such a gap – sometimes referred to as a *lacuna* – may provide your research with its rationale and justification.

- *You are interested in testing, replicating and verifying other studies and their findings in a new context* – If no gaps can be identified then it may be that your topic may be aimed at replicating some of the key ideas from the literature in other, different environments. For instance, prior studies may have been based on very small sample sizes or in a limited number of locations, perhaps with very distinctive contexts. The replication and extension of ideas is very often the way forward for many undergraduate projects and master's dissertations. Again this is a valid and justified approach and will not affect your ability to gain a high grade for your work.

Other aspects you may wish to consider at this stage include the wider context of your research, including:

- *Examining ideas presented in the so-called 'grey literature'* – In deciding what to research, you are not restricted to consulting the academic literature. You may be inspired by something you read in a report produced by consultants or a government department, whether it is a certain issue or a particular finding that grabs your attention. Your dissertation can be designed to examine such issues within an academic context or perhaps with the specific objectives of testing or corroborating data.

- *Is your research timely?* – In a similar vein, you may wish to examine a topic which is currently the subject of attention among policy makers or practitioners, perhaps because it may enhance your future employability prospects. Topics go in and out of fashion within all disciplines and subject areas. For instance, 2011 was the United Nations International Year of Forests and 2002 was the International Year of Ecotourism. While the former seems likely to capture tourism academics' attention in the future (Hall 2011c), the latter was accompanied by a reinvigoration of interest in a topic that had first come to prominence in the 1990s. Alternatively, you may wish to revisit topics from the academic literature that were popular some years ago, either to explore the continuing validity of established ideas or to examine progress over time. For example, the Tourist Area Lifecycle Model (TALC, Butler 1980) is over 30 years old now (see Butler 2006), yet it continues to have enduring appeal among undergraduates, many of whom variously use it to gauge current destination trajectories as well as its continuing relevance.

You may come across the concept of 'originality' during topic selection. For instance, as part of the assessment criteria (see Chapter 15), you may be encouraged to select an original topic. It has to be said that, although we have come across this at undergraduate level, this is somewhat uncommon and originality is instead more of a criterion at master's level and certainly so at doctoral level. Simply put, the term 'originality' has a broad meaning since your work can make a so-called 'original contribution' to the subject in a variety of ways. If you are in any doubt about whether your dissertation has to demonstrate originality and how to do this, consult the checklist in Table 3.1.

Stage 3: thinking about the practical implications

Finally, you should focus your attention on the comparative ease or difficulty of researching the topic/s you are considering. As part of this, access to information should be considered in far more detail. Furthermore, it is also worthwhile to carry out a simple audit of:

1. Your resources/skills.
2. The required tasks.

Table 3.1 Types of originality in research

- Establishing a major research approach for the first time (this is extremely rare especially at undergraduate level).

- Extending an existing study into a new set of ideas (this is more often the situation at master's and PhD level).

- Researching existing ideas from the literature into different context (this is often the case with undergraduate dissertations by, for example, using a different type of case-study area or time period).

- Taking a specific technique and applying it to a 'new' set of data.

- Developing a comparative study, for example, across different areas to test the flexibility of ideas within the literature (the scope of an undergraduate dissertation may mediate against this).

- Using new conceptual frameworks to reinvestigate an established issue.

- Developing a new technique of analysis (this may involve borrowing from other disciplines and transferring into tourism studies).

Sources: Modified from Blaxter et al (1996), Phillips and Pugh (1994).

This should also include the costs in time and money that you would incur in researching the topic (see Chapter 7 for more detail). Many students are faced with conducting dissertations in their long (i.e. summer) vacations. On one level, this might be helpful in so far as summer in many destinations coincides with periods of peak visitation and touristic activity. On the other hand, this may also be the time when you planned to get a temporary job. In such circumstances you need to consider how these conflicting demands can be reconciled. For example, there would be little point in thinking of undertaking an extensive questionnaire survey, say on a daily basis over a long time-period, if you also had to work. Under such circumstances you would need to think of having an intensive period of data collection that would also allow you to work.

Making an initial decision

With these three stages complete, you should make an assessment of the relative merits of your potential topic/s. This will look at the advantages and disadvantages as well as the possibilities and limitations. For multiple options, you may even want to conduct SWOT analyses (i.e. Strengths, Weaknesses, Opportunities, Threats) in order to arrive at a decision as to which topic is most appropriate for you. Clearly, strengths and opportunities have to outweigh weaknesses and threats (i.e. risks) to make a potential topic feasible. You can explore how this balance compares between several options to arrive at an initial decision.

If your regulations allow, you may discuss your proposed conclusions from this selection process with academic staff – possibly those who may be appointed as your advisor/s (Chapter 10) – before you commence more detailed work on your research proposal. You may be offered important feedback and suggestions, as well as external, if at this stage only initial and provisional, encouragement for your topic on which you can build.

Changing your mind

Throughout these early stages of topic selection it is quite possible that you may change your topic. This is not a major problem, and it is the case that many students change their minds due to problems with data access, after talking to their friends or perhaps more commonly after consulting academic staff.

If you are in this position, don't panic and make irrational decisions. Go back to the start of the process and work your way through your new idea/s in a logical way. Being flexible and receptive to suggestions is not a problem as long as you make a decision within the time period allowed (that is, if there is a set deadline for your dissertation proposal). Clearly, it is undesirable for you to change your topic after a first proposal has been entered because this eats into your time for conducting the research proper and hence may impose further limitations on your work.

Further refining your focus

Having identified the basis of your topic is an important first step. The next task is to refine this into a workable and researchable scheme of work. At this point, many texts move on to discuss the focusing of the topic around research questions or hypotheses (Blaxter et al 1996; Booth et al 2003).

As a practical and tested alternative we would suggest that you start by thinking about key issues you would like to examine – these may be identified in a review of the literature (see Chapter 4). From this you need to write a provisional *aim* (or even, aim*s*) of your research (see Box 3.3). Try to do this as a starting point: this will start to shape your topic and the process of constructing your *objectives* and/or any *research questions* if you elect to use the latter. Ultimately these form the basis for the development of your detailed research proposal (see Chapter 6). Sometimes this is referred to as drafting the 'research problem'.

Box 3.4 shows an example of a research aim and related objectives based on an initial idea of a topic on tourist behaviour. Aims, objectives and research questions may be distinguished as follows:

- *Aims* are general statements and give an overall direction to your work.
- *Objectives* link to the aim but provide more detailed statements of intent that can be researched.
- *Research questions* tend to be even more specific than objectives, and more importantly can very often be directly tested and subjected to variation.

Simply put, the latter are questions that are connected to, and help you deliver, your aim/s and objectives. They arise in the course of designing – and sometimes while conducting – your research (see also Box 6.3). They may be prompted by a reading of the literature and they are answered through your data collection and analysis. First and foremost, as curious scholars we tend not to think in terms of delivering objectives (like project managers or military personnel); rather, more naturally we raise (and then answer) questions that interest and excite us.

Box 3.3 Thinking through your topic selection

From the list of topics in Box 3.2, I've decided to develop ideas around the neglected area of disability and access to holidays. So far this leads me to two possible projects.

EITHER

Option 1

The study of how people with disabilities make decisions about holidays. This would involve focusing on the pre-trip stage. It could cover how promotion and marketing of holidays includes or excludes the disabled.

OR

Option 2

An examination of the meaning of holidays for people with disability. This would involve a more general study of access to holidays. It could allow me to develop a much deeper understanding of the barriers to holiday-making and -taking.

I'm going to go with the second option because it gives me greater scope to develop a topic. I could research this topic by means of a questionnaire or, better, by interviewing people. Perhaps better still might be to develop a mixed methods approach?

Source: authors

Many dissertations are restricted solely to having aims and objectives (Figure 3.1a). Others are more intricately designed and have all three (Figure 3.1b) while some eschew objectives in favour of a combination of aims and research questions (not shown in Figure 3.1). This is not our preference. Some dissertations have more than one aim and objectives connected to each. The key point is that aims, objectives and research questions have to be logically connected. Put another way, there is no point devising a research question or an objective that is not related to an aim.

In Box 3.4 you can see a clear issue (or research problem) has been identified, and this has been articulated through an aim. In turn the aim has been further teased out in terms of how it can be researched through some specific objectives. It is important to note how the aim and, more importantly, the objectives are articulated. First, the objectives are stated precisely. The use of the infinitive of the verb (to . . .) as a drafting device is especially helpful at the start of the dissertation because it indicates what is about to unfold.

A second point is that the objectives have been clarified by means of additional sentences. This is by no means obligatory or always necessary. Sometimes objectives are merely listed without such exemplification. In this case, though, helpful further details are provided on how the topic can be researched. Hints are given regarding the likely approach as well as the relevance of the objectives (which help when establishing the rationale for the work later in the write-up of your introduction – Chapter 13).

Box 3.4 Example aims and objectives for a potential dissertation topic

Aim

The aim of this research is to investigate the relationship between disability and holiday-making in the United Kingdom.

This will be researched by the following objectives:

Objectives

1. To examine the levels of participation and exclusion of people with disabilities in holiday-taking. This will research the nature and type of holidays taken by people with different types of disability as well as the relative level of exclusion.
2. To investigate the nature of holiday decision making by people with disabilities in terms of perceived and actual barriers. This will also include how problems of access impact on patterns of behaviour while on holiday.
3. To understand the meaning of holidays to people with disabilities and their families. This will draw on ideas researched with disadvantaged families. In the process, it will help to construct a more detailed picture of the importance and roles of holidays to this group.

Source: authors

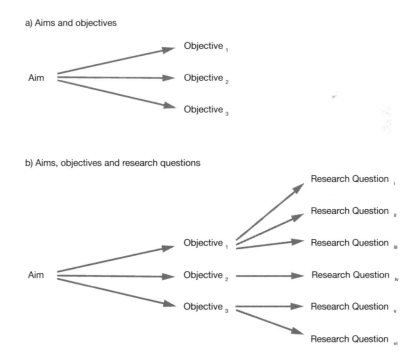

a) Aims and objectives

b) Aims, objectives and research questions

Figure 3.1 The connections between aims, objectives and research questions

Source: authors

Finally, as you will have noticed, research questions have not been employed in this example. They could have been easily added. For instance, a research question spinning out from the first objective might be: to what extent are there variations in levels of participation in holiday-taking among people with different types of disability? However, you should be aware that they are not compulsory and, while some disciplinary approaches favour their use, others do not (Chapter 6). In this case, we have chosen not to employ them.

Further specific advice in this regard is provided in Chapter 6 which deals with your research proposal. However, at this point we are concerned with flexible ideas that may be developed or revised depending on your detailed reading of the literature which we cover in the next chapter.

The chapter at a glance

The main learning points of this chapter are that:

- You should chose a topic that is of interest to you.
- Your topic should be feasible in academic and personal terms in the time period available to you.
- You should decide your probable topic based on an appraisal of several ideas.
- You should work on your ability to write a series of clear, connected and researchable aims and objectives that will allow you to research your topic successfully.

Dissertation checklist

Before you go further in your work, check you:

1.	Know what the deadline for your topic selection is.	
2.	Have a practical topic you can complete in the time available.	
3.	Have discussed your ideas with faculty members or your advisor.	
4.	Understand your aims, objectives and/or research questions.	
5.	Have a proposed research topic that fits with the assessment criteria in your institution.	

4

WHAT'S BEEN DONE BEFORE? WORKING ON YOUR LITERATURE REVIEW

Learning outcomes

By the end of this chapter you will be able to:

- Explain why your literature review is important to your dissertation.
- Plan your search for academic literature.
- Develop a strategy for what to read.
- Write with the literature review in mind.

Why read the literature?

Your research must be linked to the existing body of knowledge (see Figure 2.1). Your reading informs your work and, as we discuss in Chapter 3, it provides the initial basis for selecting a topic and developing your ideas. The ability to review academic contributions within your subject area is an integral part of the research process. Your research problem – as expressed in terms of your aim/s and objectives – will only make sense if it is framed within what has already been done in your particular area of study. Most of all, your understanding of the literature will form part of one of the assessment criteria for determining what grade you will ultimately achieve (Chapter 15).

Therefore, in practical terms, you will be reading the literature for a number of reasons (Blaxter et al 1996), including to:

- Establish the context of your topic with a view to establishing your likely contribution (perhaps originality) and putting together your proposal (see Chapter 6).
- Develop a (conceptual/theoretical) framework for your study and to help you write your literature review as a crucial component of your dissertation.
- Inform you of the way other researchers have gone about collecting information and the techniques they have used to analyze their data (see Figure 2.1; Chapters 5 and 12).

In practice clearly there are strong overlaps between these three purposes but you should keep them in mind at all stages of your reading. The purpose of this chapter is to help you organize your search of the academic literature and then to make sense of what you have found. We should stress that this chapter almost exclusively deals with searching for, and making sense of, the *academic* literature on your topic. Search and reading strategies for 'grey literature' (that may be included as background or context – see Chapter 3) are not considered here in depth, albeit there are many overlaps and similarities in process and hence many of the ideas presented below will have resonance with that sort of exercise.

Preparing for your search of the literature

For many students, conducting a literature search can be a daunting experience as they often feel overwhelmed by the sheer size of the task. There are two bits of advice worth considering at the outset.

First, the scale of your literature review should be appropriate to your requirements and circumstances. So, you need to find out the number of words you have at your disposal. To put things in context, your literature review may only be a maximum of 2,500 words in length within a dissertation of 10,000 words; that is, around 6 to 7 pages of double-spaced text, or what you might have had to submit for an undergraduate written assignment.

Thinking about the literature review in this somewhat instrumental manner has further merits. Many students do not like the unstructured nature of the literature review where they have to set the parameters of the search and the discussion. Put another way, they find comfort in predetermined coursework questions which they have to answer. Because there is no (apparent) instruction, and they are faced with a blank page, the literature review is disconcerting. It needn't be: think about devising your own 'question' to write to. How about 'critically review the relevant academic literature pertaining to the (recent) study of [insert your topic]'. This is after all precisely what you're doing!

Conducting a literature review 'from scratch' can also be troubling for many students because they are used to receiving guidance on what to read in the form of 'reading lists'. Clearly, because your topic is relatively specialized, you have to put together your own programme of reading, and this may be the first time you have been faced with this task. As a second piece of advice, then, the scale, scope and nature of your literature review will be dictated by the choice of topic in the sense that some topics have a large amount of literature while others may have little written about them. In both cases your search strategy will need to be designed to cope with whichever situation characterizes your topic.

In the case of the former, which is the more common, you need to start by identifying the key debates within the literature. Very often those are given in textbooks or review papers that journals such as *Tourism Management* and *Current Issues in Tourism* have been publishing. From these you can decide which are relevant to your topic in terms of its aim/s and objectives. This hierarchical approach is an easy way of planning your search. This is because you can add depth to the initial breadth of your understanding of the key debates by tracking down further reading from the reference lists. In turn, these 'secondary' references may provide you with further bibliographical hints and tips to follow.

In an age of digitization and online resources, this ad hoc strategy may seem somewhat 'old hat'. However, sometimes simple or more obvious keyword searches do not reveal much. This is especially the case when you are truly faced with a topic which has generated little prior literature. In this instance a slightly different approach is called for. As outlined in Box 2.1, suppose you had decided to study the significance of leadership in the context of sustainable tourism. You would find very little in standard tourism texts. Your search would need to start with key debates within the literature on leadership (within management studies or organizational behaviour, for instance). You would then need to put these in the context of work on sustainable tourism management. Again, the hierarchical approach can be applied but in this case it starts within the leadership literature, then making links across disciplinary divides to publications on sustainable tourism within tourism studies.

Reading strategies: what to read

Before starting your literature search it is important that you are clear about your strategy in terms of what to read. Above all, it is necessary to use a balance of reading, and it is useful to recall that the best dissertations make greatest use of the most recent, cutting-edge and seminal contributions in their topic area.

As work on your literature review commences, it is worth recalling that *review papers* as well as some *textbooks* can be used to help identify the key debates and strands of discussion. You may find that more focused *edited books* on specific topics give more detail on main debates. These have often been developed from conferences, symposiums and other forums where the latest research (at the time) on a topic is presented and discussed. Finally, research monographs are highly specialized texts which are routinely based on specific research projects. These give the most detailed insights and reference lists, sometimes even bibliographies on particular topics. They have counterparts in the form of unpublished PhD theses but of course they are much more accessible to you.

The most recent information tends to be available in a variety of other reading material, namely:

• *Journals* – within tourism studies these are an increasing source, and they include many e-based publications (see Table 4.1). Most, but not all, tourism journals have their articles referenced in Google Scholar, so this can be a good starting point for finding published, refereed research. You should also be aware that accepted papers 'in press' (i.e. awaiting publication in the paper version) are often posted online, and by using them you can make use of the most current and cutting-edge research.

• *E-papers* – these consist of a mixture of conference papers posted on the Internet, along with full-text PDF papers posted on blogs and other sites (for example, the personal home pages of individual academics). We would urge caution using some of this material as its reliability and accuracy cannot always be guaranteed. Quite often, it has not been through the formal peer-review process. In contrast, textbooks, and journal papers in

Table 4.1 A selection of tourism journals

Annals of Tourism Research

Tourism Management

Journal of Travel Research

Current Issues in Tourism

International Journal of Tourism Research

Journal of Hospitality and Tourism Research

Journal of Tourism Studies

Tourism Analysis

Tourism Economics

Tourism Geographies

Tourism and Hospitality Planning & Development

Tourist Studies

Information Technology and Tourism

Journal of Ecotourism

Journal of Heritage Tourism

Journal of Hospitality, Leisure, Sport & Tourism Education (JoHLSTE)

Journal of Sustainable Tourism

Journal of Sport Tourism

Journal of Teaching in Travel & Tourism

Journal of Tourism and Cultural Change

Journal of Travel & Tourism Marketing

Journal of Vacation Marketing

Tourism Culture and Communication

Tourism in Marine Environments

Asia Pacific Journal of Tourism Research

Scandinavian Journal of Hospitality and Tourism

Tourism Review International

TOURISMOS: An International Multidisciplinary Journal of Tourism

Source: adapted from New Jour (2012) and ABS (2011)

particular, go through fairly rigorous refereeing to ensure their credibility. There are several paper 'repositories' online that are available (e.g., SSRN, EconPapers). These sites are often staging grounds for scholars as they develop papers and research that may, in the future, be published in an academic journal. As with other e-papers, then, these may not be considered as refereed in the form they are presented on those sites.

While you are conducting your searches for academic literature, for instance, through search engines like Google Scholar, you may also come across other non-academic sources that help you establish the background to your study. Although the purpose of the literature review is to consider the academic literature, it makes sense at this time to collect information in the form of:

- *Official and unofficial reports* – these are widely available. While most government reports tend to be available electronically and free, by contrast most commercial companies

Box 4.1 Examples of pro-poor tourism reports available online

Author/s	Title of Report/Publisher/Location
Ashley et al (2001)	'Pro-poor tourism strategies: making tourism work for the poor', Overseas Development Institute, London, (www.odi.org.uk/resources/details.asp?id=2358&title= pro-poor-tourism-strategies-poor-review-experience)
Ashley (2005)	'Facilitating pro-poor tourism with the private sector: lessons learned from pro-poor tourism pilots in Southern Africa', Overseas Development Institute, London, (www.odi.org.uk/resources/details.asp?id=1823&title= facilitating-pro-poor-tourism-private-sector-lessons- learned-pro-poor-tourism-pilots-southern-africa)
World Bank (2006)	'Ethiopia: towards a strategy for pro-poor tourism development', World Bank, Africa Region, (http:// sitesources.worldbank.org//INETHIOPIA/Resources/ ET_Tourism_Strategy.pdf)
Ashley et al (2007)	'The role of the tourism sector in expanding economic opportunity', Overseas Development Institute, London, (www.hks.harvard.edu/m-rcbg/CRSI/publications/report_ 23_EO%20Tourism%20Final.pdf)

There are also many other web-based resources on pro-poor tourism including The Travel Foundation 'Tourism Gateway' which provides free information for tourism researchers (at www.makeholidaysgreener.org.uk/for_teachers_students_ researchers.asp#eldis) and the Pro-Poor Tourism Partnership's library (at www. propoortourism.info/Library.html)

Source: authors

usually charge high fees. In some areas of tourism the most up-to-date information tends to be only available via the Internet. Reports and case studies of work on pro-poor tourism are a good example of this (see Box 4.1).

• *Trade and national press* – once again, these are relatively varied in their coverage and most are best accessed via the Internet. The most useful websites in the UK are online newspapers (especially *The Telegraph*, *The Guardian* and *The Times*) and BBC online, which contain extensive and indexed archives of news reports. In addition to stories covered by the nationals, LexisNexis, which may be available through your library, covers regional and local publications. As with all databases, the search is often only as good as the keywords you enter (see Box 4.2).

Box 4.2 Tales of the unexpected: the results of a keyword search on a tourism disaster

The Jurassic Coast is a World Heritage Site famed for its stunning chalk cliffs and the importance of the geological record it contains. Located on the south west coast of England, the Jurassic Coast is home to several important seaside resorts and it is traversed by one of the region's main attractions, the South West Coast Path. On 18 January 2007, disaster struck. The MSC Napoli, a large container ship, cracked its hull in high seas while en route from Belgium to Portugal and shortly afterwards it was beached on the Jurassic Coast. At the time, there was predictable public debate about the time it would take to clear up, the damage that would be done to the local tourism industry, and the unwelcome consequences of day-visitors trying to salvage what they could.

A year later, in the spring of 2008, several of our students were thinking of conducting dissertation research into the impact of the Napoli disaster. As a first step, they undertook searches of various online databases for reports, commentaries and newspaper stories since the disaster as part of their background research. Getting keywords right is vital to the success of any search. But imagine their surprise when they had to wade through content about the then Miss Naples, the Italian Serie A football club and several other extraneous topics connected to a major Italian port city. Although this was at first somewhat disconcerting, eventually they took heart from the fact that their survey had established that little research had been done – or more precisely was in the public domain – on the impacts of this major environmental disaster on tourism along the East Devon coast.

Source: authors

In addition to these different reading materials there are various dictionaries, handbooks, encyclopaedias and companions for the study of tourism, which offer definitions of key terms and in some instances short discussions of key developments (Jafari 2000; Weaver 2003; Williams 2003; Lew et al 2004; Page and Connell 2011).

Finding what you need to read

Many guides on literature searches advise seeking the advice of a librarian (Blaxter et al 1996) and they point to an overwhelming range of potential sources of information (see, for example, Ridley 2008) which you could spend all your time consulting and nothing else! These days it is most likely that you will start your search on the Internet, usually after first reading an introductory or general text to get the main thrust of the arguments.

If this the case, use Google Scholar as a starting point. Unlike the main search engine of Google, this will give you more journal papers and academic books on your particular topic. Keyword searches are an obvious way to start but of course you will probably be faced with a large number of web pages of potentially interesting reading material.

From there, you can collate the list of references you find and then proceed to the website of your institution's library and search by journal. In many cases, you will end up using third-party databases such as EBESCO and JSTOR to access some journals. Moreover, your library may already work with Google Scholar to provide a live link to the actual articles you have found there. Needless to say, this can be an enormous time saver.

Some journals and other sources (such as reports) you locate may not, however, be accessible because your institution may not subscribe to them. In which case, you may be able to obtain a copy through the inter-library loans service operated by your library. Through this service, your library will attempt to acquire a copy of the reading from elsewhere through a series of reciprocal agreements. The service might attract a charge and you should factor in the time needed to get hold of particular resources as part of your planning (see Chapter 7). While it is wise to search smartly and not to rely on inter-library loans, the service can be an important supporting facility and you should make yourself aware of how it operates at your institution.

Finally in this regard, national library services may provide you with an opportunity to locate what you require. Collections such as the British Library in London, the Library of Congress in Washington and the Staatsbibliothek zu Berlin are so-called 'deposit libraries' to which recognized publishers are obliged to send one copy of everything they publish. As such, these libraries have very comprehensive holdings of material published within their countries (as well as extensive holdings of materials published elsewhere). In fact, it is the deposit status of the British Library that underpins the operation of the inter-library loans system in the UK. Such libraries are publically funded collections. Their catalogues are available online (Ridley 2008). In addition, you should be able to access them in person but you may have to apply for a reader's card to enter the facility (and you may need to have a letter of introduction from your institution and/or your student card when applying). The main advantage of using such facilities is the range of material and coverage. The disadvantages include: the cost of each trip (financially and in time); there may be restrictions placed on what you can access and/or copy; delivery from the stacks might be slower than you expect; and other users may be using what you want to read.

Producing an effective literature review: reading with a purpose

Producing an effective literature review is of course more than just collecting together a range of references. After all, your literature review is one of the central features in the architecture of your dissertation, providing a foundation for the design of your research more generally as well as for your data collection and analysis more specifically. It is though crucial not to overlook that the purpose of your literature review is to put your work into a *critical* context, not merely to describe what others have done. We will return to the idea of a critical context later in this chapter, but for now we will concentrate on the idea of reading with a purpose; that is, to explain why your work is important and how your research problem relates to the literature.

Purposive reading means that you need to decide what issues and ideas are relevant to your study. Some students make the mistake of collecting lots of literature but find that very little is of actual relevance to their own research.

Purposive reading is not an easy thing. As we have noted previously, a good starting point is to identify and list the key debates in your topic area. Then you should relate these to your specific research objectives, to see:

- If they fit; and
- How closely they relate to particular objectives.

List out the relevant ideas in terms of how strongly and accurately the links fit with your objectives and use these as a guide to the collection of references. You should look for how these relevant ideas have been debated or contested; that is, what different views exist within the academic community on these major ideas. These are vital nuances. Finally, organizing your references by theme, date of publication and, of course, relevance gives you a useful organizational framework (see also Chapter 3).

Some authors have suggested producing a so-called 'literature map', which is a linkage diagram about themes in a particular research area (Creswell 2003). This is similar to the concept of mind mapping discussed in Chapter 3. It starts with general themes and moves to more detailed and specific ones that relate to the main themes.

Healey and Healey (2003) provide a simple checklist method of organizing your literature by its relevance (see Table 4.2). This is a little basic and subjective but it does offer a simple means of evaluating the material you have collected. For instance, you should not be too hasty to dismiss papers over five years old. As we have noted above (see Chapter 3), some of the most important papers in a subject area may be older than this and there may be collections of work around key events or fashions in tourism research that require you to take a longer view of the literature.

Table 4.2 Deciding on your key references

Criterion	Possible	More doubtful	Probably forget it
Relevance to my topic	High	Moderate	Tangential
Recently published	Last 5 years	6–15 years old	Over 15 years old
Authority – the author or paper is cited in the references I have already read	Much	Recent paper not yet had time to be cited extensively	Older paper cited infrequently or not at all
Reliability of source publication	Much	Recent paper not yet had time to be cited extensively	Older paper cited infrequently or not at all
Reliability of source publication	Published in major journal	Publication is not in key tourism journal	Informal publication possibly unreliable Internet source
Nature of publication	Peer-reviewed academic journal	Textbook or conference proceedings	Popular magazine
Originality	Primary source of information – the authors obtained information using reliable and recognized methods	The authors took their information from clearly identified and reliable secondary sources	The authors provide little information on how material was obtained

Source: Modified from Healey and Healey (2003: 31)

Purposive reading also extends to how you read individual texts. When you are organizing your reading, make notes on the material you have read. There is a strong temptation merely to print off the journal papers or photocopy sections of books and think that is all you need do; however, you must also record what you have looked at, why you selected it, and make notes about the content (Creswell 2003). More specifically these notes should try to cover the following:

- A clear record of the research problem/theme the material covers.
- An indication of the ideas debated. What are the key points being debated ? Put these down in basic note form (not written out in large chunks).
- The methods used to collect the information, whether quantitative, qualitative or mixed methods approaches were employed (because these will inform what you do, as you will see in Chapters 5, 7 and 11).
- The type/s of analysis used (because these will inform what you do, as you will see in Chapter 12).
- The major findings and conclusions, and how they are recorded (because this will help in the presentation of your research, see Chapter 13).
- Any problems identified in the approach.

It is also worth noting the type of vocabulary and terminology which is employed. You may also reflect that no text is 'neutral' and that each text reflects the positionality of its author/s. In other words, it is often useful to ask yourself the following questions (which are especially helpful when reading the 'grey literature') about every text you read:

- *Who* – are the author/s and where do they come from? Is this significant?
- *What* – was their subject and what were their main findings? Why are these important?
- *Why* – was the research conducted? Why was it conducted in this manner? Who sponsored it and what influence might this have had?
- *When* – was the research conducted? When was it published? Is the date of publication significant? How does it coincide with other contributions in the same field?
- *Where* – was the research undertaken? For what reasons?
- *How* – was the data collected/analyzed? What survey instruments or analytical techniques were employed? What was the sample size? How was the sample devised?

Ridley's (2008) extensive guide to literature reviews demonstrates that your notes serve an important further purpose. They help counteract the risk of plagiarism. As she correctly notes, there are many definitions of plagiarism, but the main idea is that you should not pass off others' work as your own. She employs a definition from the University of Keele (2007, in Ridley 2008: 98) that encapsulates the full range of this form of academic malpractice:

Plagiarism is the use of the ideas, words or findings of others without acknow-ledging them as such. To plagiarise is to give the impression that the student has written, thought or discovered something that he or she has in fact borrowed from someone else without acknowledging this in an appropriate manner.

She identifies four main ways in which plagiarism may come to pass in your work (Ridley 2008: 98), namely:

- 'The wholesale copying of another's work and claiming it as your own';
- 'Rephrasing someone else's original ideas and not acknowledging the source';
- Reproducing what is considered to be common knowledge but from a published source; and
- 'Acknowledging the source but using the exact wording of the original or wording that is close to the original' without appropriate direct quotation.

By taking effective notes, Ridley argues that you are forced to articulate the content of the reading in your words, not necessarily those of the original author/s. It is easy to see how one or more of the above instances may come to pass if, for example, you are making heavy and exclusive use of photocopies or online reading. Put another way, it is useful not to compartmentalize plagiarism as an issue that arises only when you write up your work. Rather, the roots of plagiarism are often to be found much earlier, in the preparatory stages of your dissertation. As we noted above, we would encourage you to consult your own institution's guidelines or codes of practice on acceptable academic practice, and any advice we provide here is not a substitute for that. Nor do we have the scope to treat this topic extensively given the many dilemmas and debates it can raise. While we mention plagiarism briefly later, by raising the issue here we are encouraging you to start as you mean to go on, namely acting with the highest standards of integrity.

Reading for methods and analytical techniques

At the risk of stating the obvious (because it is mentioned in both the checklists above), you should not forget to read for, and makes notes on, the methods and techniques employed in extant work. It is easy to be seduced by main arguments or compelling results, or to focus predominantly on someone else's (thematic) literature review because it gives you a sense of where the debate is going and how to inform yourself further.

As we noted in Chapter 2, your review (i.e. reading of the literature) should inform your eventual choice of data-collection methods and analysis techniques, and in this regard you should not forget to read the methods sections of contributions most relevant to your topic. Above, you were encouraged to examine the general approach taken in extant studies, but you should be more curious. You can pick up some important hints, tips and benchmarks from such features as the:

- Choice of subjects, participants or (case-study) organization/s.
- Location/s for the research (i.e. the sampling frame).
- Timing of the research (i.e. when the data were collected and why).
- Sampling strategy/ies.
- Sample size/s.

- Survey instrument design (note, these are often to be found in appendices).
- Use of standard or common scales.
- Choice of analytical techniques and why they were justified.
- Key metrics or indexes employed in the reporting.
- Vocabulary of reporting.

If more than one source was employed, you should furthermore consider:

- How the choice of methods was justified.
- The sequence in which the methods were employed.
- How the data generated by each source was used (e.g. reported separately, together and triangulated, or in a complementary manner with one source informing the design of another).

Identifying common denominators: meta-analysis

These checklists offer a basis on which to devise a spreadsheet in order to record the characteristics of your reading. In the case of methods and techniques this would allow you to look for common denominators in, or trends over time related to, how a particular topic has been studied. For instance, in the third edition of a guide to qualitative inquiry and research design first published in 1998, Creswell (2012) notes a number of classifications of qualitative methods and the apparent variation of use in different disciplines and over time.

In effect, what we are describing is a technique called 'meta-analysis'. This sounds far more complex than it is and it is a form of evidence-based review (Hart 2005) or research synthesis (Cooper 2010). It is a technique that has been used widely in medical studies quite specifically to help make sense of results emanating from multiple studies on a topic (Cooper 2010). In the case of the social sciences, it simply describes a situation where you identify key, overarching features within the literature. This may include the main debates, the major findings, the types of methodological approach adopted, preferred (case-study) locations or common limitations and pitfalls.

As Table 4.3 indicates, it is a technique that has started to make its way into tourism research, and it is a device that you can utilize in your dissertation. Gallarza et al. (2002) presented a review of the main conceptual constructs and approaches to measuring destination image, following widespread interest in the topic area during the previous two decades. In the case of the latter, they used diagrams to classify and describe how quantitative and qualitative methods had been deployed on this subject. In the case of the former, not only did they classify the different types of topics covered by the various authors, but they mapped the most common attributes used in tourism destination image (TDI) studies. From Table 4.3, residents' receptiveness, landscape and surroundings, and cultural attractions were the three most-studied attributes while information, service quality, transportation and social interaction had been – relatively speaking – somewhat overlooked (among this sample of studies). This might, of itself, provide a student with a rationale to conduct research on TDI in these areas; however, such a decision should be modified based on publications in the decade since this paper appeared.

Table 4.3 Meta-analysis in action

Authors	Various activities	Landscape, surroundings	Nature	Cultural attractions	Nightlife and entertainment	Shopping facilities	Information available	Sport facilities	Transportation	Accommodation	Gastronomy	Price, value, cost	Climate	Relaxation vs Massific	Accessibility	Safety	Social interaction	Resident's receptiveness	Originality	Service quality
	Functional →																	*Psychological*		
1 Crompton (1979)								✓		✓	✓	✓	✓	✓			✓			✓
2 Goodrich (1982)		✓		✓	✓			✓		✓	✓			✓					✓	
3 Sternquist (1985)		✓		✓	✓	✓		✓		✓	✓			✓					✓	
4 Haahti (1986)		✓	✓	✓	✓			✓				✓		✓	✓				✓	✓
5 Gartner and Hunt (1987)		✓	✓					✓		✓			✓						✓	
6 Calantone et al. (1989)	✓	✓		✓	✓	✓		✓	✓			✓		✓		✓			✓	
7 Gartner (1989)		✓	✓	✓	✓			✓											✓	
8 Embacher and Buttle (1989)	✓	✓		✓							✓	✓	✓		✓		✓		✓	
9 Guthrie and Gale (1991)	✓			✓		✓	✓		✓	✓	✓		✓	✓		✓	✓			✓
10 Ahmed (1991)		✓	✓	✓	✓	✓		✓					✓						✓	
11 Chon (1992)		✓	✓	✓		✓				✓	✓	✓	✓		✓	✓			✓	✓
12 Fakeye and Crompton (1991)	✓	✓	✓	✓	✓	✓	✓	✓	✓	✓	✓	✓	✓	✓	✓				✓	
13 Crompton et al. (1992)	✓			✓		✓						✓	✓		✓			✓	✓	✓
14 Carmichael (1992)	✓												✓		✓				✓	
15 Chon (1992)	✓	✓		✓		✓		✓		✓	✓			✓	✓				✓	✓
16 Echtner and Ritchie (1993)		✓	✓	✓	✓	✓	✓	✓	✓	✓	✓	✓	✓	✓	✓	✓	✓	✓	✓	✓
17 Driscoll et al. (1994)	✓	✓		✓	✓	✓				✓	✓		✓	✓		✓	✓	✓	✓	
18 Dadgoster and Isotalo (1995)			✓	✓	✓	✓		✓		✓	✓		✓			✓				
19 Muller (1995)		✓		✓	✓	✓				✓	✓	✓	✓	✓	✓	✓	✓		✓	
20 Eizaguime and Laka (1996)					✓				✓	✓	✓	✓		✓		✓	✓		✓	
21 Schroeder (1996)	✓	✓	✓	✓	✓			✓		✓	✓	✓		✓				✓	✓	
22 Ahmed (1996)		✓	✓	✓	✓	✓		✓											✓	
23 Oppermann (1996a, 1996b)		✓		✓	✓					✓	✓	✓	✓	✓			✓			✓
24 Baloglu (1997)		✓	✓	✓	✓	✓				✓	✓	✓	✓	✓				✓	✓	
25 Baloglu and McCleary (1999)		✓		✓	✓					✓	✓	✓	✓	✓				✓		
Total	8	19	12	18	17	15	3	16	8	14	15	16	12	12	12	10	7	21	7	4

Source: Gallarza et al (2002: 63)

To attempt your own meta-analysis, as a starting point use a reliable textbook or, if available, a review article that surveys recent, cutting-edge developments or the 'state-of-the-art' in a particular field or subject. By means of tabulating data about published studies, you are able to identify where the main attention has so far been focused and conversely where there are substantial gaps in knowledge (i.e. this may provide the rationale for your study when you later write up your introduction). By entering details, for instance, on year of publication or the methods used in particular studies, you can reveal temporal or methodological patterns such as when and/or how a particular topic has been studied. The latter may provide some justification for the choice of methods you will employ and describe later in your dissertation.

Recording what you read

In order to keep a careful and consistent record of your reading material and to reference this in a consistent fashion, you should employ just *one* of the two main reference methods:

- The *Harvard System*, where the author/s name/s is recorded in the text along with the date of publication – e.g. Smith (2008). It is the system used in this book and if you turn to the reference list you will see how books, edited chapters, journal papers and websites are listed.

- *Oxford Notation*, which employs footnotes and/or endnotes. This is far less popular in the social sciences and it tends to be employed more in the arts and humanities: for example, historical studies of tourism (Semmens 2005).

You should check which reference system your institution expects you to use. Software like EndNote can help you to manage your reference list but consistency is the key. There are several variations of the Harvard System of notation and you should employ just one throughout the entire text. In the unlikely event that your institution will offer you a choice of whether to use the Harvard System or Oxford Notation, do not 'mix-and-match' or change your system of notation in different chapters of your dissertation. We would recommend the Harvard System because it is so widely used in tourism studies, and major tourism journals will offer you hints on best practice (Table 4.1) in particular for non-standard sources such as newspapers, reports or other web-based sources. No doubt, it will only be necessary to consult these in the very unlikely event that you have not been thoroughly trained in the principles and details of referencing much earlier in your studies.

Finally, you should be especially diligent in recording sources obtained from the Internet. This is perhaps more important with respect to the 'grey literature'. Many reports and semi-official publications – especially those from government departments – are only being published online (in PDF format). The Harvard System has a structured way of recording the bibliographical details of web-based sources and this includes the date of last viewing. This is because web pages and documents are often revised and go through several versions. As such, it is also wise to download and archive copies of them, where possible, so that if your work is ever called into question you have the original sources to hand. All too often, web content is 'updated' and key sources can disappear as a result.

Producing an effective literature review: writing with a purpose

As noted earlier, the scope and length of your literature review will vary depending on your institution's regulations. However, many students get concerned about how many references they should have read as a minimum, and this is a question that advisors are frequently asked.

Sadly, there is no simple answer other than to keep in mind the following factors:

1. *That your reference list should be relevant to your topic; that is, it should fit with your objectives.*

   For instance, if you are interested in stakeholders' relationships in sustainable tourism there is a far larger and more long-standing body of knowledge for you to access than if you are interested in the relationship between retailing and tourism.

2. *That you should not use just one type of reference.*

   For example, you should not use only standard textbooks as this would convey an impression of a limited range and depth of reading to your examiners. Spread your reading using the hierarchical method we mentioned earlier in the chapter.

3. *That you should not rely too heavily on single sources.*
   On many occasions we see the same name/s being cited repeatedly and almost exclusively. There may be legitimate reasons for this. Certain sources may be especially fruitful or there may be a limited literature on a topic. However, this may also be an indication that a proper search of the literature has not been conducted, and students have not developed the appropriate breadth and depth of knowledge on their topics. In a worse case, it may be considered an instance of plagiarism (Ridley 2008).

4. *If possible try to avoid quoting a secondary reference for a key, specialized one.*
   For example, if in a textbook you see a reference to a paper on something key to your topic, search out the original reference and read it. It is always possible that the author of the textbook has missed something critical to what you are researching.

You should strike a balance between having too many references or too few. This is not easy but, according to Blaxter et al (1996), you should use reference material of all types which allows you to justify and support your ideas and to make comparisons between your research findings and other research (see Chapter 13). To this we would add the use of those references that contest or debate ideas. Conversely, there is little point in using reading that is merely there to impress your examiners or that does not add to your discussion.

Weaker literature reviews frequently lack purpose, they are unfocused, or they are too descriptive. Many just list what different authors have done, and this gains little credit in the assessment process (see Chapter 15). In contrast, what you should be doing is producing a critical review of the literature. By this we mean that the review should:

- Demonstrate that you have studied the literature, and that you have a sound understanding of it and where your topic fits in.
- Give the reader a clear picture of the key debates (theoretical, conceptual, methodological) as they relate to your topic, including those ideas that are contested.
- Include a discussion of the major shortcomings of existing knowledge as they relate to your particular topic, and how might these be addressed by your research.

To be effective your literature review should provide:

- Information on what is known about the topic.
- A discussion of the key concepts and their characteristics.
- An indication of the shortcomings and limitations in existing knowledge.
- A clear indication of any missing evidence.
- Acknowledgement of the contribution your study will make in terms of where it fits.

These ideas of critical discussion are discussed further in Chapter 13.

The chapter at a glance

The main learning points of this chapter are that you:

- Should read with purpose.
- Should develop appropriate strategies for searching out and reading relevant (academic) literature to your topic.
- Must write with purpose, focusing on the clarity and quality (not quantity) of your ideas.
- Do not overlook methods in your reading in favour of theories, concepts and findings.

Dissertation checklist

Before you go further in your work, check you:

1.	Used an effective combination of reading material.	
2.	Used up-to-date reading.	
3.	Identified the key debates.	
4.	Used relevant material.	
5.	Identified significant gaps in the literature.	
6.	Used an approved referencing system throughout.	

5

METHODS AND DATA: SOME EARLY CONSIDERATIONS

Learning outcomes

By the end of this chapter you will be able to:

- Identify a range of methods and data types you may be able to use in your dissertation.
- Make appropriate linkages from your aim/s and objectives to your chosen methods.
- Appreciate the consequences of methodological decisions for the development of your research.

Methods, precedents and choices

Your dissertation will require you to collect some data in order to investigate your objectives (and research questions if you have them). This in turn calls for you to make informed decisions as to the type/s of data you need and how to collect them. Within the early development of your topic, you should ask yourself 'what is/are the most appropriate method/s of data collection for my emerging research objectives?'. We have already touched on parts of this selection process where we started to discuss the key linkages within a dissertation (Chapter 2, Figure 2.1) and how to review the literature with an eye for methods (Chapter 4).

Previous studies can go some way towards suggesting the sorts of methods you may, can or indeed should use (either alone or in combination), the types of questions you may wish to ask, and the sorts of analytical approaches you may wish to employ subsequently (Chapter 12). However, you do not have to slavishly follow what's been done before. After all, ideas change, thinking moves on, and there is regular methodological innovation (Bengry-Hall et al 2011). For instance, it was only as recently as 2007 that the *Journal of Mixed Methods Research* was launched (Tashakkori and Creswell 2007); the first book on visual research methods in tourism was published in 2011 (Rakic and Chambers); and 'virtual ethnography' (Hine 2000) or 'netnography' (Kozinets 2002) as a means of conducting ethnography on the Internet have only really been around since the turn of the millennium.

This chapter is about providing you with the information you need to make correct decisions in the early stages of your work. It will examine the issues of types of data and the methods of data collection available to you in principle. Thus, further purposes of this chapter are to prompt you to return to your prior learning on methods and techniques as well as to seek other specialist learning resources if they are needed. To reiterate, this is *not a methods book* and we do not set out to give you exhaustive advice on designing your survey instruments, executing your programme of data collection, or the analysis of the data you have collected. This chapter will provide you with a reminder of some of the more salient points about methods as they relate to forming a proposal. However, it is no substitute for, or indemnity against, the need to do more specialized, in-depth reading on research methods (i.e. in addition to reading methods in the thematic literature) while preparing your proposal and in your subsequent empirical work.

Methods mapping

You have probably already discovered this in your studies so far, but at the beginning of this chapter it is important that you recall the following general points of advice when considering your methods of data collection (based on Denscombe 2007):

- All data-collection methods have strengths and weaknesses that underlie their use. There is no one perfect method in any given context, and in this chapter we identify some of these advantages and disadvantages.
- It is often better to view alternative competing methods as not being mutually exclusive but rather as being complementary and offering support to each other. In other words, a mixed or multiple methods approach may have some merit for your research.
- The selection of your method/s depends on the degree to which they are useful and appropriate for your work: that is, in assisting you to address your aim/s and objectives (see Figure 5.1). This means not only the ability of the method to generate the data you require but also in the context of the time/resources you have available (see Chapters 3 and 7).

This latter point is especially important. You should think of aims and objectives as directly connected with methods and you can map these three elements to consolidate the linkages. As Figure 5.1a demonstrates, it is not a case of 'one objective, one method'; a particular method may be able to deliver information to help you to satisfy more than one objective (see also Table 11.1).

A similar logic applies if you are employing research questions (Figure 5.1b). Although it may be advantageous, if not efficient, for a single method to answer a set of research questions aligned to a specific objective (like method a), this is not always the case (as methods b and c indicate). Rather, you may find that methods contribute towards answering particular questions associated with different objectives.

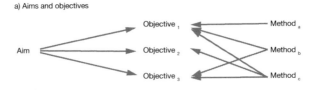

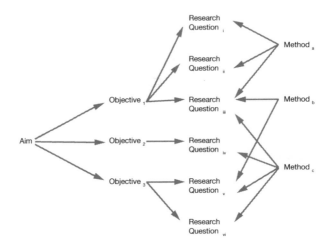

Figure 5.1 Aims, objectives and research questions linked to methods

Source: authors

The four pillars of research

Your choice of methods (and hence the formulation of your aim/s and objectives) will also depend on the nature of the research you wish to conduct. Within the social sciences there are some highly nuanced arguments relating to how knowledge is generated and presented. There are, though, four broad types of orientation you could follow, and you should check what your institution expects in your dissertation:

- *Normative research* is often characterized as research which has as its goal (or one of its goals) information and results that can be used to invoke change or guidance toward change. In some social sciences, however, normative research represents the collection and assessment of social norms, such as patterns of behaviour (Lerner 2002).
- *Explanatory* studies, on the other hand, focus on asking 'why' questions relating to particular social phenomena, and allow researchers to develop predictions and hypothesize about why things happen (Lerner 2002).
- *Descriptive* studies attempt to measure a situation or issue in as much detail as possible, thus supplying the reader with a more rounded picture.
- *Predictive* research obviously focuses on utilizing data and research findings to understand future potential and possibilities.

These four pillars are far from being mutually exclusive: in practice, it is not unusual for research to be both descriptive and explanatory, for example, and your dissertation may also take more than one approach.

As studies of tourism span several disciplines and thus incorporate multiple approaches, the range of the types of tourism research is vast. A substantial body of tourism research is inherently explanatory, largely because it employs a paradigm from social sciences that seeks to question real-world social phenomena. In addition, it is not unusual to find significant numbers of descriptive studies in tourism research. Good examples include simple visitor-profile studies that attempt to merely report on demographic and tripographic characteristics of a sample. These are valuable in their own right, but often do little to advance academic knowledge and insight. Policy studies in tourism can often be normative in their approach as they may offer assessments of future issues following present-day policy and implementation efforts. Predictive research can also be found in those studies which attempt to forecast problems or issues.

Inductive and deductive approaches

Your choice of methods *may* also relate to the method of reasoning you employ in your study. To this point in the book, we have largely assumed an inductive approach because in our experience it is the approach taken by most undergraduate dissertations in tourism studies.

Simply put, the inductive approach is based on the bottom-up generation of data which you subsequently attempt to make sense of; however, you may alternatively take a deductive approach that requires you to establish a set of ideas which you subsequently test through the generation of data in a more 'top-down' manner. Brotherton (2008: 16–18) makes several important distinctions between the two:

- *Induction* has the ultimate goal of new theory generation. The generation of data takes place early in the research process and data are derived from the 'real world' through such means as case studies (of businesses or organizations) or 'fieldwork'. Data analysis and interpretation are the early stages of a sense-making approach that requires comparisons with the existing literature and ends with the development of new constructs about the world.
- *Deduction* is based on theory-testing. In contrast, data collection takes place later in the process and it is preceded by a reading of the extant literature which results in a theoretical framework to be tested. On the basis of (working) hypotheses, a programme of empirical research is: (1) design; (2) data are collected, analyzed and interpreted; and (3) decisions are made about whether to accept, refine or reject current theory.

The choice of which method of reasoning you will adopt in your dissertation is a personal one. It will depend on your worldview and your fundamental beliefs of how research should be conducted. Conversely, it may be constrained by institutional regulations or disciplinary norms and conventions. Multiple (working) hypotheses are common characteristics of highly quantitative studies with roots in economics and psychology as well as those adhering to the principles of 'logical positivism'.

Brotherton (2008: 17) makes the very important point that you should not typecast approaches with methods in what he terms a 'deceptively simple distinction'. Often in introductory lectures inductive reasoning is associated with qualitative methods and deductive research is stereotypically connected to quantitative approaches. In the case of the former, this is because many modes of qualitative enquiry are commonly presented as being far more fluid, flexible, and iterative. In the case of the latter, the development of many questionnaires is often portrayed as part of a process that includes testing concepts or other theoretical constructs that have been identified by a review of the litera-ture. Moreover, as a means of making sense of the differences between the two, examples are invoked of deductive approaches in the form of larger-scale questionnaires being used as a means of more systematically testing constructs initially derived from smaller-scale qualitative enquiries. However, this is a false binary borne of expediency. There is no reason why qualitative research methods cannot be used to test theory, or why quantitative data cannot be used to reveal new, emergent theoretical positions.

Primary and/or secondary data

A further early consideration should be whether you are expected or elect to use primary and/or secondary data. Often dissertation students are required to collect primary data in order to address their research objectives, demonstrate a wider array of methodological skills and expertise, and hence satisfy the regulations of their programme.

Primary data refers to collecting some 'original' data for the specific purposes of your research. This may involve a range of data-collection methods including, but not restricted to, questionnaires, in-depth interviews, focus groups and researching archives. From its rawest form, it will require your processing, encoding and analysis to generate meaningful results and findings. In contrast, secondary data describes data that has been put together by another person or organization for a particular purpose but which is put to another (i.e. secondary) use in your research.

When first considering these two main types of data it is tempting for many students to make the assumption that they will just use primary data. In doing so, they very often dismiss the potential value of secondary data to their work. In many cases secondary data may help provide an initial starting point in selecting your topic (Chapter 3). For instance, a dissertation on climate change mitigation in the tourism sector may be inspired by some of the secondary data warning of the likely consequences of inaction presented in the influential *Stern Review* (Stern 2007) published by the United Kingdom government.

However, it is more likely that secondary data can provide a strong context to your specific study (Chapters 2 and 13). It may be in the form of a key finding which you may wish to examine, test or corroborate in a different setting or set of circumstances (Chapter 3). Alternatively it may present particular indices, metrics or types of calculation you may want to reproduce in your research. Other sound reasons for using secondary data in your research study are that:

- They may help fill important gaps in your primary data, particularly if your own survey was restricted in size or scope due to time constraints.
- They can sometimes be used as a comparative dataset so you can contrast your findings from your primary survey with other data.

- There may be no other means of collecting relevant data in cases where primary data collection is limited by access problems (Chapter 3).

This latter case may be applicable where your research is on businesses and commercial organizations, and therefore secondary data from consultancy reports may be the best source available.

Types of secondary data

Secondary data is available from a range of sources, but the two most frequently used are official and non-official sources. In both cases the good news is that, increasingly, this material is available electronically. This has the advantage of being relatively easily accessible.

Within the context of tourism research most countries have some official statistics on visitor numbers and patterns. However, these tend to be variable in their coverage and reliability. Table 5.1 contains a listing of some useful national websites for tourism-related statistics.

Fortunately, there are a number of international organizations that also collect together and make available tourism statistics. Two of the most important are the United Nations

Table 5.1 Examples of international and national tourism websites

International sites	Data availability
United Nations World Tourism Organisation	A range of facts and figures, mainly free together with publications
World Travel and Tourism Council	Some key international facts together with regional and national reports. Mainly freely available
European Travel Commission	Statistics and fact sheets, freely available
Organisation for Economic Co-operation and Development	Publications on range of tourism topics mainly freely available
National tourism sites	Data availability
China National Tourist Office	Range of statistics in English, freely available
Tourism Australia (student resources)	Wide range of data and information, all freely available
USA: Office of Travel and Tourism Industries (you should also check individual state tourism organizations – see example of Hawaii)	Gives general statistics and provides links to other related sites
Hawaii Tourism Authority	Range of statistics and information freely available
Korea Tourism Organisation	Statistics available
Tourism Authority of Thailand	Domestic and international statistics, availability varies for recent statistics
Caribbean Tourism Organisation (One Caribbean)	Range of statistics freely available
UK National Statistics Publication Hub	Wide range of statistics available
VisitBritain	Range of statistics available 2005–10

Source: authors

Table 5.2 Some key considerations in assessing the usefulness of secondary material

Criteria	Characteristics/concerns
Accuracy	This concerns how credible the source is and how the data were recorded. You need to examine the accuracy of all sources.
Representativeness	How representative is the material in terms of how the information was collected? Is it a sample of a population? If so, how was the sample made and how large was it? Can you check the sampling process?
Compatibility	How compatible are the data with respect to the primary data being collected in terms of scope and timing.

Source: modified from Scott (1990)

World Tourism Organisation (UNWTO) and the World Travel and Tourism Council (WTTC). In addition to such statistical data many of these organizations issue specific thematic reports which in themselves may provide you with a valuable source of secondary information for your work. The best way to discover these is to spend a little of your time browsing through the websites listed in the table.

Finally, there are commercially based sources of information which tend to be particularly useful in providing data on specific themes, locations or developments. They are especially important in giving you access to up-to-date data on issues such as the fast-moving world of e-marketing (Chapter 4). Unfortunately there can be access problems, in that, although most publications are available electronically, they are not always free to access. Many are extremely expensive and are often difficult to obtain, even in good university libraries.

As a matter of routine therefore, you should attempt to appraise documents just as you would any other academic sources (see Table 5.2). Certainly, you should be aware of any limitations in the data these sources are providing. Check for example on the sample size, how the data were obtained, and of course for what purpose and when. In other words, before using data from any source check the possible factors affecting its reliability. This is good research practice and it will improve the quality of your work.

The past and the present: documentary and archival work

When does the past begin? This is not meant to be an unnecessary philosophical question. Rather, much of the research published in books, journals and reports is inherently historical in nature. By virtue of the nature of the publication process, it is dated the moment it appears and it reports on issues that arose in the past. For instance, in recent years the WTTC (2010; 2011a; 2011b; 2012) has reported on the performance of, and predictions for, global travel and tourism during the economic downturn. These are interesting reports when read individually. When viewed collectively as a series they offer fascinating insights into how tourism leaders viewed conditions at the time, how far predictions from year to year proved accurate, and possible explanations for the variations.

Therefore it is not surprising that a great many scholars in a number of disciplines have argued that the past is the key to the present and possibly to predicting the future. In other sectors of economic activity, this has resulted in a growing literature on consumer history and business history. With a few notable exceptions, in the fast-moving world of travel and tourism historical research has not been quite as popular as charting more current trends and forecasting future behaviours. Longer-term retrospectives have great value as several studies indicate. For instance, Semmens (2005) examines how travel and tourism featured in the Third Reich to project images and further the interests of the National Socialist agenda, while Baranowski and Furlough's (2001) collection presents the rise of tourism as essentially associated with the rise of modern consumerism at the start of the previous century. Both collections make heavy use of documentary evidence from intensive archival research and they expose this as a rich seam for potential future research that currently remains largely untapped.

The trade press, minute books, advertising posters, business records, travel films, posters, postcards, photographs, personal diaries, and papers among others have the potential to be fruitful sources. Documentary and archival research can cover contemporary history as well as periods much further back in the record (Morgan and Pritchard 1999). As with other modes of research, obtaining access is a consideration, especially if the archives are held by businesses or private libraries and institutes. Moreover, archival research will test your abilities to search for, and appropriately reference, a range of historical sources.

Furthermore, data triangulation is often mentioned as a key consideration for contemporary (qualitative) sources (Decrop 1999). Nevertheless, this is perhaps more the case where the careful piecing-together of historical evidence is concerned. Even more recent database records used as secondary sources present issues of how to reconcile variations in quality, availability and comparability of data produced at different times. For instance, in a series of papers on the relationship between corporate social responsibility and financial performance among travel and tourism businesses in the United States, a group of researchers examined long-term trends from 1991 to 2006. The results are not important here. Rather, in order to be able to examine the various facets of this relationship over time, information was pieced together to ensure consistency and comparability from several standard business sources produced by a number of providers, namely: Standard & Poors (S&P) 500, Russell 1000 and Russell 2000, KLD Stats, Compustat and CRSP (Lee and Park 2009).

Quantitative and/or qualitative data

Beyond the 'origins' of the data you use in your research, you will also be faced with the decision as to what type of data you will need to collect. Here, by 'type' we really mean whether you wish to use quantitative or qualitative data (or a combination of the two) and the associated methods for collection you will need to employ.

For many students the selection tends to be based on their skills as well as their subjective preferences. For example, some students (falsely) perceive numbers as in some way scientifically more credible and, as a result, dismiss qualitative research as 'mere opinions'. Alternatively, some students dislike or are even afraid of using statistical techniques and, as a consequence, they discount collecting quantitative in favour of qualitative data. In such circumstances they look towards what they imagine is a slightly easier set of data to analyze.

There are a number of problems associated with taking a narrow or instrumental view:

- It may restrict the scope of your work.
- It is based on a false assumption that the analysis and use of qualitative data is easier than its quantitative counterparts (or vice versa).
- It ignores the apparent advantages of using so-called 'mixed method' approaches which we discuss later in Chapter 7.
- It may not be based rationally on other key aspects of research design in terms of being appropriate to an underlying theoretical approach or the precedents in the existing literature.
- Last and by no means least, contrived or narrow choices may restrict your ability to address your aim/s and objectives effectively.

In the remainder of this chapter we look at some of the basics of primary data collection by highlighting a selection of the wide array of potential methods you have at your disposal.

Quantitative methods

The use of quantitative methods to generate primary data is linked mainly to questionnaire surveys. One of the main advantages of questionnaires is their flexibility, as Ryan (1995) demonstrates. They can be used to collect basic factual information about people, such as demographic characteristics and behaviour. In addition, they can elicit data on people's opinions, attitudes and perceptions. You can also include open-ended questions which allow the respondent to express their points in their own words. These variations are a key reason why questionnaires are used so much and nowadays the routine formats are:

- *Face-to-face* – these involve the researcher making direct contact with the respondent/subject. They may be carried out in a number of environments: for example, in the street within a tourism resort or at an attraction, at people's home or holiday accommodation. Response rates vary depending on the location, weather conditions, time of day and of course the length of the questionnaire.
- *Postal* – obviously in this instance contact with respondents is limited and such a survey may be expensive to operate (Chapter 7). Response rates tend to be low (usually between 20–30 percent) but follow-up letters and reminders will improve this. Obviously the length and complexity of your questionnaire will be an important factor.
- *Internet-based* – quicker and cheaper compared with the first two but response rates are very often much lower, for example 10–20 percent. They are increasingly used by students due to the cost/time advantages and the fact that data 'entry' is also quicker, allowing for more timely data analysis. Response rates depend on layout, design and length of the questionnaire as well as the type of delivery (Dillman 2007; see also Kaplowitz et al. 2004). Mailed-out postcards inviting people to fill in a web-based survey have been used as a strategy to increase response rates.

Web-based technology has expanded the potential horizons of data collection, and web surveys/questionnaires can now be relatively easily managed by researchers using their own domains. Firms such as Survey Monkey (www.surveymonkey.com) can also be used to collect raw data which can then be easily imported into statistical software packages. One issue to be aware of, however, is the extent to which particular segments of the population and groups in society have access to the Internet in their homes.

Some software now allows questionnaires to be designed as 'apps' for mobile phones and tablets, with the intention of increasing response rates. Therefore, in addition to directing respondents to a website to fill in a questionnaire, visitor studies can make use of mobile technology on-site at major tourist attractions or information centres, where tourists can be invited to spend five or ten minutes filling in an online questionnaire.

Obviously each survey method has advantages and disadvantages. You need to appraise the alternatives with respect to their ability to deliver timely and relevant data of sufficient quality. You should also appraise your ICT knowledge and skills. There is nothing worse than a survey that generates responses only for the data to disappear into the electronic ether (see Box 14.1).

Qualitative methods

The use of qualitative methods has become increasingly popular in student dissertations. Such approaches cover a variety of methods including in-depth interviews, focus groups, participant observation and even diary-based methods (Phillimore and Goodson 2004). Other newer, more innovative methods employing visual data in the form of photographs, video diaries and drawing have started to emerge (Rakic and Chambers 2011). However, they have some common characteristics as identified by Rossman and Rallis (1999) and Creswell (2003), which include:

- The use of multiple methods which are both interactive between the researcher and respondent, and humanistic.
- Such methods are not tightly constrained within a study but rather emergent; that is to say, they may evolve during your research as you learn more about what questions to ask. As Creswell (2003: 182) contends, 'the theory or general pattern of understanding will emerge'.
- Qualitative methods are useful for researching complex ideas, especially opinions and perceptions.
- Research involving such methods may take place in the natural setting of the respondent (Rossman and Rallis 1999). In tourism studies this natural setting may include holiday accommodation, a bar in a holiday resort, or even a bus or aeroplane for a highly choreographed package trip (Stephenson 2002).
- Such research raises important ethical issues (see Chapter 9).
- The methods are interpretive, which puts an emphasis on you as the researcher to interpret and make sense of the data. This has implications on how you present and analyze the data (Chapter 12) along with the fact that the data is filtered through you personally. You need to take note and acknowledge possible biases such as your

Box 5.1 What do you see? Multiple readings of a heritage tourism site

A piece of art? A powerful memorial? Just another part of the 'concrete jungle'? The answer will depend on your background and worldview.

Let's get the basic facts out of the way first. In the foreground of this photograph is the Memorial to the Murdered Jews of Europe, to give it its formal title. It was designed by the architect Peter Eisenman and inaugurated in 2005. Its geographical coordinates are 52.5139°N, 13.3783°E and it is located in the Mitte district of Berlin.

For a great many this is a powerful memorial that commemorates the murder of many thousands of Jews during the Third Reich, a space of acknowledgement and atonement, a site of dark tourism close to other symbolic sites such as the Brandenburg Gate. For others, it is also an impressive achievement of engineering and architecture that extends the boundaries of design for memorials. As a result, it has almost become a 'must see' for visitors to Berlin (although there weren't many visitors at the time the photograph was taken). Those with a keen historical eye may see a monument oddly in keeping with the surrounding urban heritage in the government district of the city with its rich, multiple layers of history that make Berlin such a popular city-break destination.

Concrete Plattenbau (prefabricated apartment blocks) in the background were a hallmark of post-war reconstruction in the German Democratic Republic and they are a legacy of one of the contested pasts that are so heavily commodified in the city's heritage tourism products. Finally, those interested in 'barrier free' tourism may see a site that, from this angle at least, does not at first sight appear especially accessible to people with mobility impairments.

Source: authors

positionality (e.g. age, gender, ethnicity, education, background). As Box 5.1 indicates, there are a number of different lenses through which to examine the world.

- There is an inherent contestability in qualitative data collection and analysis requiring greater reflexivity or reflection of/on interpretation (Alvesson and Sköldberg 2009). This forces researchers to consider moving between different theories, concepts or positions to arrive at a critical interpretation of the data.

To elaborate this final point, Alvesson (2011: 6) has argued with respect to interviews that the complexity and uncertainty of the practice of research makes the meaning pulled from them contestable. Instead, he advocates a 'reflexive pragmatism' in which 'a commitment to accomplishing a result' (i.e. 'making "strong" knowledge claims') is balanced against the completion of 'alternative lines of interpretation and vocabularies and reinterpreting the favoured line/s of understanding through the systematic drawing upon any alternative points of departure'. The latter may be based, for example, on the theories or positions one holds (Alvesson 2011: 7).

When selecting qualitative approaches some students merely state they will 'do some interviews', but in practical terms there is a much wider range of methods to choose from, both in terms of data collection and data analysis, and moreover there is greater methodological innovation (Silvermann 2011). Weaker students also commonly have the misperception that they are able to conduct interviews because they know how 'to have a chat'. It perhaps goes without saying but there is far more to sound and effective qualitative research than this.

Types of interview methods

There is a variety of interview methods, but before we discuss these it is important that we remind you to think carefully about why you think interview data is of use to your project. You should also keep in mind that interviews are not the 'soft' or 'easy' option; they require a good deal of planning and preparation to be properly conducted (see Chapters 7 and 11).

The main types of interview methods are:

- *Unstructured interviews*, which tend to follow a general set of ideas/topics to be discussed with the respondent, but which may be asked in no particular order, and some may not be covered at all. Your role as interviewer is not to be too intrusive: just start the interview and let the interviewee develop their ideas.
- *Semi-structured interviews*, in which you produce a clear list of topics/questions but are flexible as to when and how these are discussed within the interview.
- *Structured interviews* are more guided by you, the interviewer. The interviewee is asked a set number of questions usually in an organized sequence. Remember, the aim of all interviews is to allow the interviewee to give their views expressed in their own words. These interviews obviously work best in a face-to-face situation but of course in many cases this is not possible and the interview may have to be conducted by phone, e-mail or Skype.

When to use focus groups

Increasingly, methods training includes the use of focus groups (also called 'discussion groups') and these have been increasingly popular in tourism research. The main advantages of focus groups are that they bring together a number of people in order to study their group norms, meanings and process (Bloor et al 2001). As Barbour (2008: 133) emphasizes, their role is to 'uncover *why* people think as they do' (emphasis original) and they are particularly effective in revealing 'how views are created and modified through group interaction'.

A focus group consists of a small number of people, usually between about six and eight, and a moderator. This is the person who will introduce various topics of discussion to the group and guide the discussion based on a topic guide (i.e. script): in other words, you!

Focus groups are not intended as a forum to survey the views of a number of individuals in one place at a given time (i.e. group interviewing) and they are unlikely to be successful if they are used in this way (Bloor et al 2001; Barbour 2008). Instead, they offer the means to establish consensus and dissonant positions over key issues, and how these form. The dynamics of the group can often lead to unexpected outcomes or new directions for the research, and the social relations of how positions are negotiated and reconfigured are often as rich a source of data as the final outcomes. This is why, for many years, it was common practice for market research companies to video groups (notwithstanding the ethical considerations this required).

Successful focus groups depend on the interaction of the group and the skills of the moderator in facilitating a productive discussion on topics of interest. If you haven't been trained in how to run a programme of focus groups, do not assume they are a straightforward technique. If you haven't had this as part of your taught modules, you may consider going on training courses (perhaps outside your institution) with all the implications this entails for your timetable (Chapter 7) and in terms of cost.

Like interviews, focus groups are *not* just a case of sitting a few people down in a room, presenting a topic and letting them chat among themselves. They require careful design, they can be time consuming to operate (see Chapter 7), and they can entail significant costs (Bloor et al 2001; Barbour 2008). You need to recruit people, find a venue, and encourage people to participate – often you will be asking them to give up between one and two hours of their time. These time problems are usually overcome by paying people to attend (i.e. incentives and travel expenses) but obviously this is costly. For many student dissertations, it is the time and cost of focus groups that tend to restrict their use.

Therefore, you need to ask yourself 'Do I need to conduct focus group research?', and 'What is the added value of assembling a group of people versus a series of interviews singularly with each participant?'. Focus groups are useful under the following circumstances (Denscombe 2007: 180):

- When you are interested in collecting underlying motives and reasons that help explain specific views.
- In exploring particular topics when you want to obtain a broad feel for how different people consider them.

- For considering the degree to which people agree and share views in terms of your research topics. This is sometimes used to help in the early stages of questionnaire design.
- For presenting tangible objects and artefacts (e.g. brochures, postcards, plans, mock-ups) on which you want the group's opinions.

Most focus groups are conducted face-to-face, but there are circumstances when they are conducted virtually using VOIP (voice over Internet protocol). These are difficult to set up but can be undertaken via 'bulletin boards', chat rooms and other social media forums. One further disadvantage is that it is more difficult to observe, note and record the group interaction.

The use of observation methods

Data can be obtained by observation: you can watch and record people's behaviour and listen to their discussions. Such observations can give you a direct record of exactly what people do: not what they say or think they do, but rather their actual behaviour. Of course, this form of empirical observation does not, on its own, give you any information on the motives behind a particular type of behaviour. To obtain this, observation has to be conducted in conjunction with other forms of qualitative research methods, typically interviewing.

There are two types of observation: participant observation and systematic observation. Participant observation is when you as the researcher infiltrate a particular situation. For example, taking a package tour to observe the behaviour of other tourists. Obviously there are problems of access, either through costs (i.e. paying to go on a tour) or through needing permission. There are also clear limitations, particularly age – as a younger person you would be in a difficult situation in joining a group of retired people. There are also issues of your personal safety (Chapter 8) and ethics (Chapter 9). Before deciding on or using this approach you should be aware of the following:

- Age and gender can act as key limitations.
- Your ability to blend in and observe are of key significance.
- It is not an easy approach since it demands a great deal of effort and commitment from you, the researcher.

Systematic observation covers the situation where it is difficult for the researcher to participate. In basic terms it refers to gathering information as an observer but without any participation. Usually you start with a list of items you are keen to observe and record in a systematic fashion. The approach is therefore based on a methodical listing of items that you observe from patterns of behaviour. As a researcher you need to consider the practicalities of undertaking this method, such as:

- Locating yourself to be unobtrusive and avoid any interaction with those being observed.
- Having a complete view of the entire sample under observation.

- Having a means of recording events with you such as a notebook, clipboard and/or camera. As such, you will need to work hard at being part of the background.

As a final note on observation, you should be aware that, just as is the case with other methods, you need to consider the ethical implications and take care to ensure that you obtain approval for your work (Chapter 9). As part of this, you should consider whether there are any legal impediments that may preclude observation and/or the recording of activities as well as the potential risks to your health and safety (Chapter 8). In some instances, photographs or videos may be prohibited (unless you have special permission). There may even be issues with copyright, particularly if you are observing or recording a commercial activity. The owners of this may even claim ownership of your notes or recordings on the basis that commercial confidentiality may be at risk if the results of your work are made public. Another potential issue arises when you observe illegal behaviour during the course of your fieldwork but subsequently fail to report this to the appropriate authorities. It is critical, therefore, that you familiarize yourself fully with the local regulations and legal responsibilities that govern the conduct of your research. Your advisor/s will be an obvious source of guidance on these matters, if not directly then certainly as a liaison with ethics officers (Chapter 9).

Diary-based methods

This approach involves information being collected by respondents asked to keep a diary and record within it specific information relevant to your research. For instance, they can be used to generate rich data on time spent on particular activities and at specific attractions or destinations. They can also be used to record experiences, emotions or the details about how travel decisions were made.

In one sense, then, diary-based methods may represent an extension of the interview process. The result is a form of extensive narrative which must be read carefully and in context. Like interviewing, diary-based research also requires a good deal of preparation, it can be demanding of your time, and there are significant cost considerations. You are asking people to keep a record of their behaviour while they are at leisure so you will need a high degree of cooperation. To achieve this first you will probably have to pay people to complete the diary.

You should only consider using a diary-based approach if:

- You really need such detailed information and you have discounted alternative methods.
- You are able to get cooperation from the respondents.
- You have the time and resources to collect such information.
- You are able to cope with the analysis.

The chapter at a glance

The main learning points of this chapter are to:

- Think carefully about what data you are going to collect.
- Consider how your aim/s and objectives, as well the type of research you wish to conduct, drive your choice of method/s and data type/s.
- Reflect on the implications of your choice of method/s and source/s for your ability to address your aim/s and objectives.
- Reflect on how your choices are influenced by precedents in previously published studies.

Dissertation checklist

Before you go further in your work, check you:

1.	Know whether your regulations require you to use (exclusively) quantitative or qualitative methods or a combination of methods.	
2.	Know for sure that you are required to undertake primary empirical research or whether secondary data will suffice.	
3.	Are aware of the types of methods used by previous studies related to your topic.	
4.	Understand the merits and limitations of employing specific methods, in particular as they relate to your topic.	
5.	Have chosen an appropriate array of methods and sources to be able to address your aim/s and objectives.	

Part II
PROPOSAL

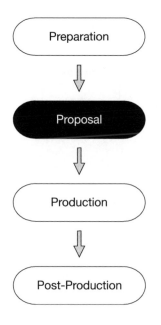

6

FORMULATING YOUR RESEARCH PROPOSAL

Learning outcomes

By the end of this chapter you will be able to:

- Identify the components of a research proposal.
- Explain why it is important to invest time in properly articulating your plans.
- Write clear and robust objectives for your research.
- Assemble a comprehensive and successful research proposal.

Torment or opportunity?

Preparation is crucial to every successful dissertation. This is perhaps never as important as in the run-up to, and the submission of, your research proposal. This is effectively a summary of your thinking so far. It is your chance to inspire confidence among your future advisor/s in your ability to conduct your research independently and successfully.

Rather than view the dissertation proposal as a 'necessary evil' you are compelled to complete, you should consider it to be a great opportunity to enhance your project and build your confidence. Put another way, it is a chance to obtain vital feedback on your planned research, to clarify particular issues, and to validate your work. In some cases, it is trickier to hold on to this bigger picture. For instance, your proposal may be directly assessed and contribute (summatively) in part to the final grade for your dissertation.

The central message of this chapter is that you should not skimp in your efforts to produce as full and systematic a dissertation proposal as possible. This is irrespective of whether your dissertation proposal is formally or informally assessed: in our experience, there is a direct link between the standard of the proposal, on the one hand, and the quality and performance of the final dissertation on the other.

To be absolutely clear, skimping on your dissertation proposal is a false economy: if you do not take the dissertation proposal seriously, you may be making a potentially grave error by introducing unnecessary risks into your project. Depending on your programme regulations, you may not be permitted to proceed to independent research until you have completed a satisfactory proposal. In this case, you may have to resubmit successive documents until you reach an acceptable standard. Long after your initial deadline has passed, your precious time towards the final submission deadline may be ticking apparently all the faster with your progress stalled at the proposal stage (see Chapter 7).

Even if you were (or have been) allowed to proceed with an incomplete or weak proposal, this is no cause for celebration. Eventually you will encounter the same types of issues that you should have considered earlier but, more worryingly, this is likely to be in the course of conducting your research when you may feel more stressed. As such, you may not be fully prepared or in the right state of mind to follow the best pathways or 'firefight' problems you could – and indeed should – have anticipated and avoided. In the worse case, as a result you may take a series of inappropriate decisions that result in flawed work. Therefore, it is far better that you invest a reasonable proportion of your total study time for the dissertation in establishing and anticipating how the project should work.

Your proposal revolves around your aim/s and objectives (Chapter 3), how they connect to the reading you have undertaken (Chapter 4), and how they drive your choice of approach, methods (Chapter 5) and analytical techniques (Chapter 12). Thus, in the next section of this chapter we build on our previous advice by offering some further guidance on how to write efficient and actionable objectives. This is followed by a discussion of the main components that usually feature in a dissertation proposal and their role/s. The final section of the chapter considers what makes a good dissertation proposal by presenting you with a series of questions you should pose yourself when you are putting this document together.

We start from the assumption that you have decided on a single topic area, narrowed down from several initial ideas (Chapter 3), and that you may have already done some preliminary reading of the connected thematic literature (Chapter 4), and perhaps also in the area of research methods (Chapter 5). To be clear, it is not necessary for you to have completed this reading fully by the time that your proposal is submitted. Reading is an ongoing process throughout your research. You may have asked the views of academic staff on your (initial) ideas but we assume – even advocate – that advisors have *not* been appointed yet (although we recognize that in some institutions they may have been). This is because the proposal represents a moment of commitment where it is vital for you to buy-in fully to your project (not theirs).

Reinforcing connections and continuity

By having your aims and objectives firmly established before you commence your empirical work, you maximize the chances of engineering high levels of connectivity and continuity throughout your research, culminating in your final dissertation document. Before you start to compile the narrative of your proposal it is important that you cement the direction your research is going to take from this point onwards. Many other decisions which need to be encapsulated in your proposal build on these foundations.

In Chapter 3, we set out the basic differences between aims, objectives and research questions, with a view to prompting you to think about these in the early stages of topic formulation. An aim is a much more general statement of ambition for your dissertation. Effectively it sets out the broad purpose of the research. Of course, your research may have more than one purpose, in which case more than one aim is justified. Objectives are more tightly focused statements of intent, setting out more precisely the direction of travel for your study. In consultancy and contract research, they are often called 'terms of reference' because they establish the boundaries and parameters of a project.

While many students find it quite straightforward to recognize these differences, they find drafting these statements an altogether trickier task. This is because it requires a clear focus and, above all, the intention to commit on a direction for their work. You may have a pretty clear concept 'in your head', but putting this down on paper can be awkward. Try to capture what it is you wish to achieve in your research in a sentence of no more than 20 words (per aim).

Somewhat predictably, the next step is to consider what follows on from your aim/s; that is to say, what specifically you propose to address in more detail (i.e. your objectives). To do this, try writing a further series of short statements that elaborate your aim/s. Precision is vital, and it is also the most significant challenge in this exercise. When you are drafting objectives try to use tight, careful wording and specific (even technical) vocabulary.

Precisely how you represent your aim/s and objectives is a matter of personal preference. Some students prefer straight text. You might want to adopt our approach of using a kind of chart or map to make links (see Figure 3.1); alternatively, you might prefer to use a 'spider diagram' or another device with which you are familiar.

Whatever your choice, you should recall that aims and objectives have to be connected. Objectives are a means of ensuring that the aim/s of your study are addressed. It is, therefore, illogical if your objectives do not offer the means by which to address your aim/s.

You should give some thought to the approaches, methods and even the types of data you will require to satisfy your objectives. As we have already noted (Chapter 5), your choice of methods is connected to, not to say driven by, the intentions for your study (see Figure 5.1). There is no point collecting or presenting data that has no purpose and hence by implication is unessential. Start to make basic connections between your objectives (via any research questions) with the possible methods you are going to use to address them. If you are feeling confident, you may also decide to elaborate the connections further; that is, from your objectives (and via research questions and methods) to the types of techniques or procedures you are going to use to analyze your data (Chapter 12).

By now your piece of paper might be quite full! In parallel, then, you may want to prepare a second document, explaining:

- Why you have chosen these aim/s and objectives (as well as research questions and hypotheses if you are employing them).
- Why you have drafted them in this particular manner. For instance, why have you used particular (subject-specific) language or (technical) vocabulary?
- Their justification or importance in the (academic and/or grey) literature (i.e. the rationale behind their choice).

- The sort/s of data you require in order to be able to address or answer them.
- The type/s of analytical techniques you may need to employ in order to address (or answer) them.

This document might even take the form of a table. However, by keeping these explanations separate from a simpler (diagrammatic) articulation of your aim/s, objectives and methods, you will keep the latter clearer while you have more space to elaborate the former.

If you haven't already spotted this, such a document is very useful in three respects:

- First, it serves as part of the records of why you made particular decisions during the research process which will be extremely valuable in writing up (Chapter 13). Remember, your dissertation is a measure of your skills as an independent research worker, one crucial part of which is your ability to make appropriate decisions.
- Second, in compiling this document you are establishing the linkages between the different parts (i.e. future chapters) of your dissertation for yourself. In Figure 2.1 we have presented the principle in a more abstract sense; through this exercise, you are engaging in the practice 'for real'.
- Finally, you may have realized that as a consequence of drafting your objectives you will have to consider ethical issues (Chapter 9) and how you will conduct the work safely (Chapter 8).

Developing your terms of reference further

As you start to work your way through this task, it becomes progressively easier because your thought processes will gather momentum. Drafting one objective may induce a series of conducted issues (i.e. objectives) and/or (research) questions you would like to consider and, in turn, you'll naturally start to think about how you are going to research them.

In many cases, we have advised students at this point to articulate as many objectives (and research questions) as they can. The point is, it is easier to consider ideas during the design stages of your research (and then subsequently discount them) rather than to add objectives, for instance after you have commenced your programme of empirical work.

Of course, what can result is an extensive – and wholly impractical – 'wish list' of things that students would ideally like to do during their dissertation research. Therefore, when you have your aims and objectives drafted the next step is to appraise their relevance. Each objective should follows from your aim/s. If it does not, the objective should be either deleted or revised so that it does. Not only are superfluous objectives damaging to the continuity of your study, but also they have an 'opportunity cost' in drawing resources away from other aspects of your research if they are not spotted and eradicated early.

If your regulations permit you to discuss your ideas with academic staff before submission of your proposal, these discussions are also likely to be far more productive if you take along an early draft of your aim/s and objectives: your advisor/s will develop a clearer idea of the direction your thinking is taking you, and you will receive more focused feedback and suggestions from which you can develop a more thorough and compelling dissertation proposal. For instance, you may receive advice on the appropriate number and wording of objectives for your topic.

One of the common questions we are asked by students is 'How many aims and objectives should my research have?'. Of course, there is no single, unequivocal answer to this question. The usual response is that there is no 'magic' or perfect number. All projects are different, and the number of aims and objectives will vary from project to project depending on the nature and scope of the research a student is proposing to conduct.

Somewhat understandably, many students think that we are 'sitting on the fence' when we say this but it is in fact true. More complicated studies may have more aims and multiple objectives. It does not necessarily follow, though, that longer or larger-scale projects have to have more aims and objectives.

In fact, we are great believers in simplicity in research design. Remember, the more objectives (and research questions) that you have, the more you have to address adequately in your analysis and discuss in your findings (see Chapters 12 and 13). Clearly then, the more complex your aims and objectives, the greater the risk that there may be the sorts of inconsistencies we have already identified as the hallmarks of weaker dissertations (Chapter 2). For instance, you may neglect the consideration of one objective entirely in the analysis or only present a partial analysis with respect to another because your project is too ambitious and you have left yourself too little time.

Perhaps the best universal advice we can offer is to keep your aim (notice the use of the singular) and objectives as simple, clearly articulated and directed as possible. A single aim can be the most efficient and effective orchestration for your study when well written.

Where possible, we try to exclude the use of 'and' and 'or' from objectives (they should not really need to be used in an aim either). Both these little conjunctions may unsettle your readers, they serve to widen the scope of objectives (when a narrower focus is preferable), and they are, simply put, unnecessary. Both force your readers to make assumptions, and remember that 'assume' can be short for making an 'ass of (yo)u and me'! Consider the following:

- In the case of 'and', you are articulating two aspirations. It may be a little passé, but why not write two shorter, simpler objectives instead?
- 'Or' introduces a degree of confusion and doubt. If you observe, record and analyze a particular phenomenon related to such an objective, at the completion of your research which part/s of the objective have you actually addressed? When you have your data, can you be sure you will know precisely what will it be telling you?

You can introduce further precision by employing best practice advice in strategy and planning, namely that objectives should be 'SMART' (Murphy and Murphy 2004: 103):

- Specific (clear, straightforward and directed to a particular goal).
- Measurable (so that you can gauge your progress).
- Attainable (appropriate to the task at hand, with an achievable scope).
- Relevant (applicable to the current operating conditions or parameters).
- Timely (with a clear start and end point).

Using this sort of framework is useful in so far as it forces you to think about the practicalities of your research (see Chapter 7). You should consider all aspects for each objective. You may write a specific and timely objective which is not attainable in terms of scope. There is no point in having grand designs that are unrealistic within the time you have at your disposal.

Box 6.1 Some examples of poorer objectives

I   To examine the extent to which there are different behaviours and motivations related to food tourism among people from different social classes and nationalities

II   To assess the importance of events to the development of urban tourism in London.

III   To investigate the meaning of the Edinburgh comedy festival for performers and policy makers.

IV   To explore the images/expectations that American students had of Mexico before they travelled there.

V   To consider the future importance of the Internet for holiday promotion by major European tour operators.

Source: authors

Most of the objectives in Box 6.1 contravene the basic rules of simplicity and specificity. Objective I is not only vague (i.e. Which people?) but also far too complex. Behaviour and motivation are two aspects of the tourist experience that are often discussed in tandem, but they are two quite discrete aspects so why not have objectives that deal with each one individually? Arguably, there are four potential objectives within this statement based on the permutations or, put another way, a potential dissertation on its own. Objective II is also vague (i.e. Which events and over what period/s?) and unspecific (i.e. Development in what respect?). What is more, the scope of the task is potentially massive. London – or for that matter any global city – hosts so many events and has such a complex tourism sector that it prompts the question of whether it is really possible to isolate the effect particular events can have. Objective III is misguided because there is no logical reason why performers and policy makers should be grouped together, while Objective IV falsely equates images as equivalent to expectations. Images may help mediate expectations but they are not the same. Even if the '/' is meant to be shorthand for 'or', it is not clear why a student traveller should have one or the other.

Finally, Objective V is an example of where a student has tried to develop a policy related or applied objective, with a view to culminating the dissertation with a forward look based on the previous analysis. There is nothing wrong with this laudable aspiration; however, as drafted this objective is not well thought out. 'Consider' is a somewhat general

---

Box 6.2 Some examples of stronger objectives

I   To explore the ways in which disabled people negotiate the pre-trip stages of the tourist experience.

II To establish the level of wilderness attachment among visitors to the Yosemite National Park.

III To investigate the ways in which small- and medium-sized accommodation providers in rural Otago make use of the Internet in their businesses.

IV To evaluate the importance of film and TV as a motivating factor for visitors to the Yorkshire Dales.

V   To assess, using Hall and Sharples' (2003) typology, the extent to which we can talk of gastronomic tourism to Padstow (Cornwall, England).

Source: authors

---

word which is at odds with 'importance' which is more definite, although it's not clear how this will be judged. Moreover, 'promotion' in what regard? Perhaps this means marketing or bringing destinations to market? Does it entail advertising, taking enquiries, or online sales? The problem is, we can't be clear.

Box 6.2 present some examples of stronger objectives. There is no doubt that each would benefit from the sort of exemplification discussed in Chapter 3. The first objective makes clear which group it is addressing, at what point in the tourist experience, and the process involved (i.e. the discussion). Objectives II and III are really quite specific and narrow, and they leave little room for misinterpretation. Both define where the research is going to be conducted and what it will focus on. Objective IV does include 'and', but quite deliberately so. Yorkshire, as so many other destinations around the world, has attempted to market itself based on its fictional portrayal in the visual media. Film and TV are connected if not synonymous (i.e. films appear on TV), many people do not appreciate the difference between the two media, and the two are used to co-brand the destination, so it makes sense to consider them as a package.

Objective V is an example of where a student wants to employ a specific framework – in this case a typology of food-motivated visitor – as a lens through which to observe a phenomenon. Padstow is famous as the home of the internationally renowned celebrity chef Rick Stein, and anecdotal evidence suggests that people are motivated to visit Padstow for short breaks or day visits just to visit his establishments. Hence, the student clearly wanted to assess the extent to which gastronomes (i.e. those motivated strongly by food as the peak visitor experience) feature among the visitor profile in what would otherwise be just another Cornish coastal resort. Finally, there is a variant of this type of objective and it is to examine the veracity of a particular set of ideas. Thus, another version of this might have been 'to assess the relevance of Hall and Sharples' (2003) typology as a means of categorizing food-related visitors to Padstow'.

Research questions and hypotheses

Research questions are even more specific articulations of your aim/s and highly directed elaborations of your objectives. Research questions may be inspired by your basic intellectual curiosity or from something you have read in the literature. In one sense, then, they may be indicative of a more inductive approach to research (Chapter 5). However, as Box 6.3 indicates, it is not a great jump in logic to progress from research questions to using a more structured set of hypotheses which you would test; that is, a more deductive approach. Irrespective of the method of reasoning you propose to employ in your research, now would be a good time to draft these devices if you are going to use them, and to note why they are written in a particular way.

Like strong objectives, good research questions should be clear, precise and narrowly focused, in the process servicing the aim of the research (see Box 6.3). The best research

---

Box 6.3 Aim ... objective ... research questions ... hypothesis

An aim of . . .

'To examine the nature of the wilderness experience in a popular, high-profile national park'

is accompanied by an objective of . . .

'To establish the level of wilderness attachment among visitors to the Yosemite National Park'

which, in turn, raises several specific research questions about the level of wilderness attachment, including . . .

- Are there gender differences?
- Are there age-related differences?
- Are there differences between day-visitors and those staying overnight?

the first of which may prompt the following hypotheses . . .

H_0 There is *no* significant difference in the level of wilderness attachment between male and female visitors to the park (null hypothesis).

H_1 There *is* a significant difference in the level of wilderness attachment between male and female visitors to the park (alternative hypothesis)

which may be examined by means of a quantitative approach employing a questionnaire survey. However, difference – or not, as the case may be – may also be revealed by the responses to a programme of interviews (although such a formal articulation of hypotheses in this manner is unusual in qualitative research).

Source: authors

questions will suggest the necessary source/s and analytical technique/s required to answer them. To be clear, research questions used in the research design are not necessarily identical to questions as articulated in your survey instruments (Chapter 11). Several questions within a questionnaire or semi-structured interview schedule, for instance, may be required to answer a particular research question.

Not all students choose to use research questions to shape their work *and* their final document; some only use them while they are conducting their research (i.e. 'behind the scenes' and not in the final write-up). However, irrespective of how or when they are deployed in the research process, their real virtue is that they force you to provide answers and hence directly address your objectives (in the final text). In the process, they should minimize the risk of dislocation by ensuring that what you said you were going to do is in fact what you did in the end (Chapter 2).

If it is your choice as to whether or not to employ hypotheses, you should consider their relative merits. For their critics, the expression of (a series of) hypotheses can be somewhat tedious, disrupt the early flow of the text, and result in a somewhat contrived or artificially segmented commentary in the later chapters of dissertations. Their advocates argue that, if constructed properly, hypotheses should make links to, be informed by and demonstrate your understanding of, background reading. Moreover, they can contribute to enhanced continuity within the text, if all of the hypotheses are addressed and reported in the data analysis and conclusion.

As you may recall from your methods training or reading, inherent within deductive approaches is the formulation of hypotheses and the use of inferential statistics, or a series of statistical procedures designed to allow the drawing of an inference (i.e. conclusion) from the data you have collected. More specifically, they relate to how confident you can be that you are making valid inferences from your sample data. Probability levels – otherwise known as 'confidence levels' – are used (e.g. 95 percent, 99 percent, 99.9 percent) to indicate just how certain you can be about generalizing from your results (Brotherton 2008).

Inferential statistics demand a particular approach to drafting hypotheses. A null hypothesis is accompanied by an alternative hypothesis. A null hypothesis merely states that there is no significant difference between phenomena, while the alternative hypothesis asserts there is (see Box 6.3). This is tested through a so-called 'two-tailed' or non-directional test. If a clear type of difference is perceived (i.e. A > B or A < B), then the alternative hypothesis has to be drafted to reflect where the difference may lie and a so-called 'one-tail' test is employed.

Further operating parameters

While we have concentrated on aims and objectives as vital aspects of your preparation, there are two other basic operating parameters for you to consider at this stage: time and location. When and where are you going to conduct your research? What is more, you must justify these decisions in your proposal first of all and later in your dissertation.

As we have noted in previous chapters, your background reading will go some way towards suggesting where your research should be conducted. Other considerations may be your curiosity, wanderlust or the simple matter of cost (i.e. the un/availability of funds).

However, as Box 6.4 indicates, a more detailed tactical assessment of the precise location/s may be required after you have broadly identified where to conduct your work.

In this example, Greece is identified as a highly appropriate country in which to work on the issue of young people as cultural heritage tourists. The problem is, there is almost an embarrassment of riches because there are so many famous archaeological sites. In this case, the student narrowed down the choice from the macro (i.e. country) to the micro (i.e. Delphi) by means of an evidence-based view. A framework was compiled and a final decision made on the balance of the evidence (i.e. critical reasoning was employed). A further round of decision making, which is not described in the box, could have been applied as to precisely where at the site the student would undertake the surveying (e.g. in the car park, at a particular monument, by a visitor centre, at a kiosk). You should remember that each of these decisions has the potential to impact on the nature and volume of information you are able to gather.

In the next chapter we examine the need to plan and manage your use of time as a key resource in some detail. However, more basic considerations are when in the calendar you are going to conduct your data collection and whether you are going to use one or more episodes. You may feel under pressure to get started as soon as possible, to make the most of the time available. However, once again a more considered approach should yield better dividends. You should consider that the timing of your data collection has to be appropriate to the basic method of understanding we identified in Chapter 5. It also has to be appropriate to your topic and subject/s, as Box 6.5 indicates.

Tourism products and offers are often highly constrained in time and space. For instance, events, festivals, celebrations and anniversaries form the bedrock of many tourism marketing campaigns, but they take place intermittently and at particular spots. So, if you want to study them you have to be in the right place at the right time. There may be more general reasons for identifying broad windows for your research. For example, you may wish to study visitors to tourist resorts in high season, out-of-season or in the so-called 'shoulder months' that form unique start-of- and end-of-high season periods. To do this, you will need to understand the nature of the season and dynamics of the market in your study locations, for instance, when businesses periodically shutdown (Duval 2004) or move to tap different seasonal markets (Blumberg 2008).

By conducting your background research properly, you will be able to determine when to undertake each part of your work and when you will have to concentrate your efforts (i.e. spend more time than the average). In the case of surveying mountain bikers, the student had the luxury of conducting research throughout the final year, but this may not be the same for all students and can create the need for some careful thinking. For instance, if a student in the British system were required to concentrate on the dissertation in the final semester, this might mean that the majority of the hours devoted to the module would be programmed in from January/February to March/April with hand-in perhaps in May, June at the latest. Clearly this presents significant issues in terms of access to, and the potential profile of, visitors because the shoulder months are traditionally around Easter and peak season from June to August.

Box 6.4 Beware of Greece bearing research gifts? Choosing a study location

Cultural heritage tourism has become a popular topic in recent times and the power of the past has become a lure to many students. In an acute observation, one student noticed that there were very few studies of young people as tourists and their motives for, and experiences of, heritage sites.

In order to address this apparent gap in the literature, the student, who lived in the country for several years and was fluent in the language, decided to conduct research on the topic in Greece during the Easter vacation. Where better than Greece with many sites of classical antiquity? Delphi, Delos, Mycenae, Olympia, Knossos to name but a few. And therein lay the problem: at which site/s to conduct the fieldwork?

No recommendation had been made in the research proposal which had, instead, just set out a series of potential options. After asking the advisors' views, it was decided to narrow down the research to a single location. This was for reasons of cost because the student was self-funding the fieldwork and two or more sites would be too expensive as well as impractical in the time available. Furthermore, comparison was unnecessary. The research was exploratory in nature and a case-study approach was being taken. No claims about wider generalization would be possible.

A 'long-list' of eight sites was drawn up. The student claimed that no detailed market research was readily available that included definitive data on the age profiles of visitors to these sites. The advisors were in no position to contest the veracity of this observation. Instead, the student was advised to take an alternative approach to decision making by devising a matrix. Into this data on several diagnostic features were entered where they were available and/or could be derived, among which were: the number of visitors in total, domestically and internationally (because one of the objectives was to make comparisons between the two); visitors per day; number of car parks; nature of the visitor facilities, the ratings and entries in popular mass-market guidebooks used by young people, and characteristics of the site.

The aim was to identify which sites may appeal most to young people as well as deliver suitable opportunities to interact with sufficient visitors to generate a robust sample of questionnaires and interviews. A combination of quantitative and qualitative data were entered in the matrix. This confounded the use of a simple scoring system to derive a single score to determine site choice. Instead, the matrix was used to structure a discussion with, and the student's recommendations to, the advisors at the next progress meeting. In the end Delphi was chosen which seemed strangely appropriate in view of its history.

Box 6.5 Creatures of habit? Timing your sampling

Mountain bikers are, it seems, creatures of habit. There are clear patterns to when they use trails. Nevertheless, many students overlook the fact that that a sample can be a function of, or reflect the conditions at, the time it is drawn.

There are, for instance, often great variations in visitor numbers and profiles in high versus low season. Moreover, while you may employ the principles of random sample generation on visitors to an attraction such as a theme park or museum, there is still a distinct possibility that if you draw a sample in the school holidays it may be different to one generated in term time. This is a potential limitation to your research which you may wish to point out in your dissertation, particularly if you are only able to sample at one point in the year.

Clearly then, the time at which you draw your sample is as important to your sampling strategy as the mode of selecting individuals to participate. For this reason, and if time allows, you may be able to widen the period in which you sample. One of our students was interested in researching the value of mountain biking to the visitor economy by means of a questionnaire. In short, local tourism marketing had emphasized outdoor experiences and encouraged mountain bikers to visit the area, but were they coming and how much were they really contributing?

With the opportunity to conduct fieldwork throughout the final year and a purpose-built trail on the doorstep, the student was able to pick up 'trail count' data; that is, every time a biker passed by, it was recorded digitally by trail managers. On the basis of daily counts, weekly and monthly use profiles were compiled. These revealed how to spread data collection around the year in order to generate as representative a sample as possible (anecdotal evidence suggested there were differences in types of visitor based on season – i.e more 'fair weather' riders in the summer). Moreover, it indicated the relative balance needed between weekday and weekend.

Indeed, the data were so rich that they were able to suggest the times of day that were most commonly and heavily frequented at the site. Armed with these data and with a strategy in place, the student went out to interview with a view to drawing a random sample and claimed that everybody who completed the trail had an equal opportunity to complete the questionnaire. The reality was, though, that a convenience sample was generated, with all those people who completed the trail having the opportunity to participate in the survey when the student was on-site.

The moral of the story? You should aspire to the highest standards of practice in conducting your research but it is almost inevitable that you will have to make compromises along the way. It is important to explain and justify how and why such decisions were made and the possible impacts they have on your work.

Key components in your dissertation proposal

After you have settled on your basic building blocks, it is time to develop the rest of your proposal. You may be required to complete a form on which you answer a series of standard questions. The importance your programme attaches to particular issues will be indicated in how much you are expected to write or, as in Box 6.6, the number of marks afforded to each section in an articulated marking scheme. Your ability to answer these questions competently and in appropriate detail will form the basis of the assessment of your proposal.

Alternatively, there may be more flexible instructions. You may be required to complete a proposal based on several much broader themes but within a prescribed word limit. In this case, you must decide on the relative importance of the issues as they relate to your topic.

Irrespective of the precise system used in your programme, you should seek to ensure that in preparing your dissertation proposal you consider the following:

Aim/s and objectives Often one of the shorter sections of the proposal, you are required to establish what you are trying to achieve through your research.

Context/background This section of the proposal establishes the background to, and rationale – or defence – for, your research (Chapters 2 and 3). It explains how your study compares with, or differs from, other existing (academic and/or practitioner) studies on, or connected to, your topic area (Chapter 4). In so doing, it establishes the potential originality and contribution of your research to tourism studies. Even undergraduates can – and indeed should – seek to make a contribution (however modest) to the body of knowledge on tourism. If you are conducting case-study research, you should attempt to describe, explain and justify your choice of destination/s or business/es based on their relevant framing conditions. You should also briefly explain the timing of the research.

Selection of methods and analytical techniques Your choice of data sources (primary and/or secondary, qualitative and/or quantitative) and analytical techniques should be thoroughly described and properly justified with respect to your aim/s and objectives, as well as precedents in previously published academic and practitioner studies. You should set out the main operational issues associated with the conduct of your data collection. The central consideration here is whether your strategies for data collection and analysis will generate adequate material to satisfy your aim/s and objectives. At the proposal stage, you may wish to include – perhaps as an appendix – a SWOT (Strengths, Weaknesses, Opportunities, Threats) analysis. You should also outline your sample frame/s, your approach/es to sampling, and the size of the sample/s you expect to draw.

Draft survey instruments These may not be required in your proposal and if you are not confident about your topic, you may be unwilling to invest time in early development work. Still, if the purpose of the exercise is to obtain feedback as you prepare to embark on your research proper, you may wish to include early drafts of your survey instruments (i.e. questionnaires, interview schedules, focus-group topic guides etc.) for comments. These are usually included as appendices.

Box 6.6 A possible articulated marking scheme for a dissertation proposal

Criterion/indicative questions	Marks
Aim(s) and objectives • Have the overall aim and research objectives been clearly specified? • Are they appropriately explained?	/15
Relevant literature • Are the aim(s) and objectives located in the relevant academic reading? • Is the research informed by relevant theory and concept? • Has the academic rationale for the research been fully explained?	/15
Background • Have appropriate non-academic sources been consulted? • Has the non-academic rationale for the research been fully explained?	/10
Methods, sources and analysis • Are the choices of research methods appropriate? • Has the methodological approach been adequately explained? • Is it appropriate to the aim(s) and objectives? • Are the choices of analytical techniques appropriate? • Is the rationale for case-study locations or businesses properly explained?	/30
Project management • How comprehensive is the project plan? • Is the scope of the research feasible in the time available? • What are the health and safety issues involved in conducting the research? • Are appropriate measures suggested to address them?	/15
Ethical considerations • What are the ethical issues associated with the research? • Are these fully described? • Are appropriate measures suggested to address the ethical issues raised by the research?	/15
Overall Mark:	/100

Source: authors

'Project management' You will be required to identify the main tasks involved in the dissertation, and when they will be undertaken in the time available. Although it may not be necessary to do a full project management plan in the strictest sense (see Burke 2003), a Gantt Chart or Critical Path Analysis (see Chapter 7) may be a sensible way of illustrating the sequence of tasks, their connections and overlaps within your work. This information will also go a long way towards convincing academic staff (i.e. faculty) of the feasibility of your completing the work in the time available, as well as your plans in the events of certain contingencies (see Chapter 7).

Ethics As discussed in Chapter 9, almost inevitably your research will raise some ethical issues. Although ethical considerations are perhaps most intricate, complex and challenging with respect to qualitative methods, they are raised by quantitative approaches, too. Your proposal should seek to identify the ethical issues raised by your research and explain how you will address them. As part of this, you may wish or be required to include drafts of letters requesting access or informed consent forms (as appendices).

Health and safety Like ethics, health and safety is often a set of considerations that is overlooked or underplayed in many dissertation proposals (see Chapter 8). All types of research have implications for your health and safety to one degree or another. Your health and safety statement should seek to identify the hazards and assess the risks involved, with plans to manage the risks you identify to be at unacceptably high levels.

References Your dissertation proposal is, like anything else you produce for your degree, a piece of academic work subject to the conventions of your institution (see Chapter 14). These will, no doubt, include statements about plagiarism and you should act to avoid any accusation of this. You will have used extant academic and other sources to generate ideas, shape your aim/s and objectives, and contextualize your forthcoming research. A simple insurance is to reference properly throughout your proposal and to produce a consolidated list of references at the end of the text. Check you know the referencing system used at your institution, as well as whether the references contribute towards the final word count of your proposal: for example, is it 2,500 words including, or exclusive of, the references? Lists of references can consume large numbers of words; the good news is that they are rarely included in the word count.

Appendices These are a useful way of including additional supporting material while keeping the main body of the proposal succinct. Once more, appendices tend not to be included in the word count for proposals. So, there is the opportunity to demonstrate that you have (more) fully considered the research by the inclusion of greater supporting materials. Don't go over the top, though: it may be that the more material you include in the appendices, the less inclined they may be to read it all.

And finally, you may be asked to sign a declaration that you have read and understand your institution's regulations on plagiarism. By the final year of your studies, we are sure that you must have encountered, and been intensively informed of the requirements as they relate to, this very important issue. Please refer back to your institutional guidelines and consult your dissertation handbook to make sure you are fully up to date on this matter. Suffice to say that, as the dissertation is intended to be the pinnacle of your programme, now is the time to ensure that you conduct yourself with the highest standard of academic integrity.

What makes a good dissertation proposal?

The simple answer is a proposal that incorporates all of the above with a high level of execution! Many dissertations programmes use a standardized pro forma to evaluate proposals like the hypothetical example presented in Box 6.6. If this is the case, you will probably be issued with a copy of this when you start your dissertation work.

The precise constitution and appearance of these evaluation forms is not as important as the issues they are likely to interrogate. Several indicative questions are listed below that are routinely used in one guise or another to assess dissertation proposals. Not all of these questions (or variations) will be used by every tourism programme. In addition, as we noted above, the weightings attached to – and hence expectations of – particular aspects of the proposal will vary. Therefore, rather than work 'blind', you should find out which particular questions drive the assessment of your work, and which aspects are more heavily weighted. The questions you should be considering include:

- Are the aim/s and objectives clearly articulated?
- Are the aim/s and objectives clearly located in the relevant academic reading?
- Is the dissertation informed by appropriate theory and concept?
- What is the rationale for the research?
- What is the background to the research?
- Are the choices of research method/s and data source/s appropriate?
- Are the choices of analytical technique/s and data presentation appropriate?
- Are there any practical issues surrounding access to, or availability of, data?
- When will the research be conducted and is the timing appropriate?
- Where will the research be conducted and is the choice of (case-study) location/s appropriate?
- Is the scope of the project feasible for the time and/or financial and/or human resources available?
- Are the timelines for the research realistic?
- Has a project (management) plan been included?
- How comprehensive is the project plan and are there any significant missing components?
- Are there financial resources available to be able to conduct the research as described?
- What are the ethical issues associated with the research, and are these fully described?
- Are there appropriate measures in place to address any ethical issues the research may raise?
- What are the health and safety issues involved in conducting the research?

- Are the hazards fully audited?
- Are the risks properly assessed?
- Are there appropriate measures in place to manage the health and safety risks?

Assessment can be formative (i.e. contributing to your development as a researcher but without direct input to your final mark) or summative (i.e. again contributing to your development but by benchmarking your performance with a grade that contributes directly to your final mark).

In many respects, these are useful questions to ask yourself even if your proposal will be only formatively assessed. Indeed, if your proposal is summatively assessed and contributes towards your final dissertation mark (see Chapter 15) almost more important than the grade in terms of moving your research forward is the 'qualitative feedback' you will receive from your advisor/s when you meet to discuss your work (see Chapter 10).

One final point to note is that if you are to be assigned a grade, there may be a double jeopardy for poor proposals; that is, although you may be asked to improve certain aspects of your work before being allowed to proceed, sadly you won't have the opportunity to be regraded.

In the next three chapters we consider three operational matters in detail which, although crucial to the successful conduct of independent research, tend to be overlooked by a great many undergraduate students and which are either at best only partially included in their proposals or at worse omitted altogether.

The chapter at a glance

The main learning points of this chapter are that:

- **The submission and assessment of your research proposal is the culmination of the design phase of your dissertation.**
- **You should not skimp in your efforts to produce a competent dissertation proposal.**
- **You will need to invest a reasonable proportion of the total time available to your dissertation research in preparing your research proposal.**
- **A successful research proposal is one that demonstrates a full appreciation of the implications of the various practicalities connected with addressing a clearly articulated set of aims and objectives.**

Dissertation checklist

Before you go further in your work, check you:

1.	Know how your research proposal is going to be assessed and whether the assessment will count towards your final mark.	
2.	Are aware of the preferred length and format for your proposal.	
3.	Allocate yourself enough time to complete a comprehensive research proposal without cutting corners.	
4.	Understand the risks and likely consequences of putting together a weak research proposal.	
5.	Are able to write clear, focused objectives.	
6.	Know how to link your objectives (research questions and hypotheses) to methods.	

7

DEVELOPING YOUR RESEARCH PROGRAMME

Learning outcomes

By the end of this chapter you will be able to:

▦ **Better plan the time available to your research.**
▦ **Estimate the time required to conduct each part of your work.**
▦ **Manage the demands of your dissertation and other commitments.**
▦ **Identify possible sources of delay to your research.**

Time as a key dimension

Time is major consideration in the design and execution of your dissertation. It is also a finite resource. You have a limited number of study hours for the completion of your dissertation. Although time is very important to you, many dissertation proposals we review include general or sketchy timetables for the completion of the work. In some cases, proposals do not include a timetable at all!

For a prospective advisor or assessor, this hardly inspires confidence and it undermines the credibility of the proposal when it is reviewed (Chapter 6), perhaps for summative assessment (Chapter 15). Common reasons for such an oversight are that students either have not thought the issues through or that they do not wish to make themselves 'hostages to fortune'. Better to keep their options open early at the proposal stage – in due course it will become clear to them when and how much time they will have to invest in their research.

You need to plan the use of your time wisely and fully, and preferably before you submit your dissertation proposal. If you plan properly, and approach your timetable in a disciplined way, you can reduce the risks of producing a poor dissertation. By setting yourself a series of realistic time-delimited objectives (Chapter 6), you will create a very useful set of indicators against which you can benchmark your current progress. The key word is 'realistic'. Of course, you can be ambitious but your goals have to be feasible. Otherwise you will fall behind your self-imposed targets and your morale may suffer as a result. In this chapter, we follow the same approach that we have throughout the book: we encourage you to break down your dissertation into its principal components, to budget your time according to each task, and to give careful consideration as to when each component is going to take place. A sensible approach to using your time will result in a more feasible project, a less stressful research experience and, all things being equal, a better chance of a more successful result.

Initial considerations

Before you start to budget and schedule your time, there are two initial questions for you to answer:

- How much time am I expected to invest in my dissertation?
- How many weeks or months have I got to complete my dissertation?

You may think that these are one and the same question, just worded differently, but they are not.

The former refers to the total time necessary for you to invest in order to complete a satisfactory dissertation. These days, with academic programmes more tightly regulated than ever before, your institution will have determined how many hours they think you should invest in order to conduct work of pass standard or above. The amount of time varies from programme to programme and it depends on the number of credits allocated to the dissertation.

By finding out the figure, you will give yourself a clearer understanding of the expectations for your work and hence what constitutes an acceptable standard for a bare pass (see Chapter 15). The higher the number of credits and hours needed, in general the higher the standard that will be expected of you.

The latter question refers to the time frame (or window) in which you must conduct your research; that is, the period between start date and submission when you must undertake the total study hours for your dissertation. Again, this depends on your programme:

- Some programmes require their students to conduct their dissertation research throughout the duration of their final year. In effect, you may be working on your dissertation in an almost part-time manner, on and off throughout the entire year. In parallel, you may have lectures, seminars, tutorials and presentations, even course-work or examinations for other modules as you complete the requisite number of credits for your degree.
- Conversely, some programmes require the dissertation to be conducted once all other taught elements have been concluded. In this instance, you will be working on your dissertation effectively full-time with no other academic commitments to worry about. We encountered some of the issues associated with this in the previous chapter.

To recap, the answers to these questions are important because they help you to:

- Understand more clearly what is expected of you and by when (Chapter 15).
- Define the scope and scale of your research (Chapter 5).
- Schedule the timing of your activities carefully.
- Ultimately reduce the risk of producing a poorer piece of work.

Your 'average commitment'

A good starting point is to work out the average number of hours per week you are expected to invest in your dissertation from start to finish. Some programmes may even point this out in the handbooks or supporting materials you are given. However, it is important to consider what this means in practical terms for you, your studies, your personal life and any paid employment you may have. It will help you understand when to try to schedule your work as well as determine the scope of your research.

For example, let's assume that you are required to conduct 300 hours of study over three months in order to complete your dissertation. This means 100 hours per month. This isn't very helpful because it doesn't really mean much to you. On a weekly basis, however, this equates to around 25 hours per week, which is a lot easier to understand because it means roughly five hours a day for a five-day working week which most students seem to have. You can perhaps work in one longer period or work in smaller spells each day depending on how you work best.

Another way to contextualize this is to consider that a working week is around 37.5 hours (for a 9.00 a.m. to 5.30 p.m. job, with breaks factored in). So, two-thirds of your working week is going to be devoted to your dissertation (with the remaining time allocated to any other commitments left in your programme).

Of course, the length and timing of the working week varies around the world as work practices vary across cultures (Scherle and Coles 2008). So, you can adjust this basic 'rule of thumb' calculation to suit your specific needs. You should also consider that this average suggests the level of commitment when spread across the whole period of your dissertation. In fact, you are unlikely to invest the same (average) number of hours week-in, week-out. Some weeks you will work more hours and others you will work fewer. For instance, you may frontload your work to spend more time on your literature review (Chapter 4) and/or the preparation and implementation of your survey instruments (Chapter 11). Alternatively, you may wish to backload your work if your objectives are concentrated more towards analysis and discussion of the findings (Chapter 12). There may be times associated with data collection (i.e. at events) when you have to spend more time on your research.

Monitoring your progress

By knowing the average number of hours per week you are expected to work on your dissertation, you are ideally placed to plan your activities and monitor your progress. Notwithstanding the different rhythms and variations of timing in an individual project, you will be better able to assess whether you are falling behind, or are ahead of schedule, as well as whether you need to be concerned about the time you have been, and will be, able to devote to your dissertation before the submission deadline.

Time, scale and scope

Of course, your ambitions for your research have to match the time you have available. There is no point in designing a very large-scale or elaborate programme of empirical research if you have comparatively little time available in which to conduct it.

Guidelines exist in both quantitative and qualitative research on acceptable sample sizes in order to be able to collect sufficient data and apply appropriate analytical techniques to produce relevant results and meaningful findings (Krejcie and Morgan 1970; Israel 1992; Ryan 1995; Barbour 2008; Stapleton 2010). In parallel to these decisions, the selection of approach (i.e. exclusively quantitative, exclusively qualitative or mixed methods), the precise method/s, the nature of your survey instruments and the scale of your sampling all need to be relevant to the time you have available.

Time can also drive decisions with respect to the selection of your topic and your literature to review (Chapters 3 and 4). Some topics in tourism studies have been the subject of more extensive academic treatment than others, such as sustainable tourism or small- and medium-sized tourism enterprises, and they have generated large-scale extant bodies of literature (Thomas et al 2011). In order to review the relevant literature for a dissertation in these topics, you will need to spend more time than you would for less well-developed areas such as innovation in the tourism sector (Hall and Williams 2008; Coles et al 2009; Shaw and Williams 2009).

Budgeting your time more precisely

To use your time optimally, you will have to make a series of informed decisions, even some compromises about what to do, when and for how long. We have seen basic project timetables that allocate time on a chapter-by-chapter basis, from chapter one to chapter n, as it were.

While this approach has certain basic merits, a far stronger approach is to think through the time you are going to need in order to undertake the following key tasks which more closely reflect the real experience of working on a dissertation:

- Selecting your topic.
- Reading the relevant literature.
- Searching for relevant background information.
- Producing and having your dissertation proposal reviewed.
- Getting clearance for ethics, health and safety.
- Reflecting on feedback.
- Reviewing/critiquing the relevant literature.
- Designing and piloting your survey instrument/s.
- Conducting your data collection (i.e. 'fieldwork').
- Data entry, analysis and interpretation.

- Writing up.
- Compilation of tables, graphics, figures and appendices.
- Binding and submission.
- Slippage, creep and contingencies (i.e. insurance time if things don't work out as planned).

Initially, in order to address these concerns you may think of your time in terms of weeks. As you progress and reflect on your plans in more detail, you may start to allocate your time in terms of days, and later even hours. This is of course a far more precise (and hopefully more realistic) means by which to plan your work.

There are no straightforward or universal answers as to how long each of these activities will take to complete to your satisfaction. This depends on issues such as the nature of your topic as well as your expectations. However, it is useful to apply some assumptions. For example, it may be useful to think about the time it will take you to write up the final text. As a rule of thumb, some students work at a rate of around 1,000 words per day to compile their final document. Barring any mishaps (e.g. losing data, not saving work – see Chapter 14), this means that it is going to take ten working days (or thereabouts) to complete a 10,000-word dissertation. In other words, two weeks without working at the weekends or a little under two weeks if you work at weekends too. If you can work at double this pace, then – in theory at least – you may be able to put the entire final draft together in little over a week.

When you are planning your work, the precise position and sequence of these actions depends on such considerations as your preferences, the nature of the topic, and the length and demands of the dissertation proposal.

Some students and their advisors prefer them to undertake extensive literature reviews and background searches before they submit their dissertation proposals. These form the basis for the 'literature review' and 'introduction' chapters when the work is written up (Chapter 13). Some even work extensively on developing survey instruments and identifying methodological issues. This forms a substantial part of the methods chapter when the work is finally written up (Chapter 13). Taken together, this progress means that comparatively fewer tasks (mostly associated with data collection and analysis) remain to be completed once the full proposal has been reviewed and approved. Conversely, where briefer proposals are required, these may be informed by major academic thinking, 'headlines' and 'highlights' ahead of more extensive programmes of reading later on.

Some students factor the drafting of their proposal into their timetable. This is a quite sensible move. After all, in Chapter 6 we cautioned you against rushing headstrong into working on your research without a strong foundation. Some students and advisors even prefer to go through several stages of revising and reviewing drafts of proposals before the advisors are prepared to grant their students full independence to continue their dissertation work. Clearly, this diverts time away from other actions although it provides a comprehensive basis for the remaining work by ensuring that there are very clear parameters for other tasks, such as data collection and analysis.

Finally, we would strongly recommend that you include time for slippage, (mission) creep and contingencies. This is often missing from student time-budgets. For all your planning, like all other research workers you will not be immune to unforeseen events

---

Box 7.1 Immaculate timing? We can't always predict when events happen!

Skiing is a popular pastime around the world. Once associated almost exclusively with mountainous regions of Europe, ski-resorts have developed in parts of the world that are not always immediately associated with winter sports. Japan is no exception. It is a vibrant and growing destination for ski-tourism, with over 600 ski-resorts to choose from.

Japanese resorts may not be as high profile on the global stage or as well researched as Whistler-Blackcomb, but this – and their number – raises interesting research questions, not least whether the same or similar processes have under-pinned their development. And it is precisely these questions which one of our students recently started to research. With a review of the literature completed, work turned to the development of a mixed methods approach including questionnaires and semi-structured interviews. Case-study locations were evaluated and selected; risk assessments were conducted; ethical clearance was obtained; survey instruments were piloted, refined and confirmed; permissions were received; interviews were arranged; and considerable time and expense was invested in booking accom-modation and travel.

The only problem was that the date of departure was scheduled for shortly after 11 March 2011, the day the devastating tsunami hit Japan. To make things worse, several of the proposed fieldwork sites were located within the exclusion zone around the Fukushima nuclear plant. Of course, the research could be rescheduled notwithstanding lost deposits, the costs of insurance excesses, and delays to the project which could be mitigated by extension. More important than any of this is the safety of our students. More sobering thoughts are that such an extreme event is one that might not have appeared on a risk assessment, and what if our student was in Japan when this happened?

Source: authors

---

or able to anticipate every contingency that could affect your work. Abraham Lincoln, the celebrated US President once said: 'I claim not to have controlled events, but confess plainly that events have controlled me.'

Personal illness, paid employment, unsuccessful pilot episodes, poor weather conditions during data collection as well as events, crises and catastrophes can all occur during your period of research and impact on the quality and timing of your work (see Box 7.1 and Chapter 8). Progress can sometimes be slower than expected and deadlines can slip as a consequence (although you may be able to apply for an extension for delays or mitigation resulting from conditions beyond your control). In some instances, initial results or newly discovered techniques can convince you to widen the scope of your work which incrementally creeps into a larger project than you first imagined. Whatever the reason, you should plan for, and allow time to deal with, unanticipated contingencies.

Your data: your most significant investment of time

The advent of online search engines has enabled students to locate literature in their field with relative ease and in real time (Chapter 4). Gone are the days of manual index searches and ordering books from the stacks which used to take hours if not days. As a result you should be reasonably well placed to assess the time it will take you to produce a critical review of the literature (Chapters 4 and 13).

In stark contrast, many students significantly underestimate the time required to conduct their data collection and data analysis. It is not quite as simple as to suggest that work towards your literature review, methods and results chapters should take, *pro rata*, broadly a third of your time. As you will see, data collection and analysis do, however, routinely take up a disproportionately high share of the total time available to you, more if you don't plan properly or if certain events intervene.

Thus, it is vital that you budget for your data collection and analysis time as systematically, carefully and accurately as possible. If you underplay these in the planning process, you run the risk of introducing significant limitations into your work. After all, there is an old adage in computer science that is relevant to all forms of research: GIGO – garbage in, garbage out! This means that, in order to get good quality output at the end of the process, you have to input good quality information in the first place. Moreover, your data is another key means of delivering the continuity from introduction to conclusion which is a hallmark of a strong dissertation.

The first step in making more realistic assessments of the time needed to collect and analyze your data is to list the main actions you will have to conduct. These include, but are not limited to:

- Design of survey instrument/s.
- Pilot of survey instrument/s.
- The data-collection episode/s.
- Data coding.
- Data entry.
- Quality checks.
- Data analysis to generate results.
- Discussion of the findings by connecting back to the literature.

The next step is to estimate the amount of time required for each of these stages based on the answers to the following type of questions:

- How long does it take to complete one survey (e.g. questionnaires or interviews)?
- How many surveys (e.g. questionnaires or interviews) can be conducted per day?
- How long does it take to complete the processing (i.e. coding of the data)?
- How long does it take to enter the data by keystroking or transcription?
- How long does it take to post-process the data for data input errors or (coding) consistency?
- How long does it take you to complete the running of analytical tests or procedures on the data?

'Days' should be the principal unit because it requires greater precision whereas the use of (working) 'weeks' suggests a more provisional and approximate estimate.

It may not be possible for you to answer these questions with 100 percent accuracy or certainty, but you should be able to make a series of informed assumptions. These are clearly far better than not bothering at all! Three main sources are likely to be helpful:

1. Your past experience of conducting research, but since it is improbable that you have extensive experience this is probably a somewhat unlikely or inaccurate reference.
2. Your knowledge of, and training in, preparing survey instruments, part of which may have included timing how long it took to pilot and conduct surveys (questionnaires or interviews) as well as to undertake the processing (i.e. coding or transcription) of, and generation of results from, the raw data.
3. So-called 'analogue studies' from academic and practitioner circles. These are studies using the same sort of methods, survey instruments and analytical techniques with similar purposes, length and scope to your research. Within the methods sections of research articles, reports, doctoral theses and the like, there can sometimes be important clues as to how long your survey instruments may take to develop, pilot and implement.

This last point is especially relevant if one of your objectives is to emulate or to duplicate a method used elsewhere. As your context of application will almost certainly be different, analogue studies will only offer you rough estimates of the time you will need.

To assist you in your planning, you can produce a (spreadsheet) table for each method with each of these tasks as rows and the time taken to conduct them as a second column. As Table 7.1 indicates, a further column records the assumptions that you are making – for example, 25 questionnaires entered per day, or for every hour of tape three hours of transcription. You may even add a fourth column to record broadly when you propose to conduct the actions either based on calendar month or within the dissertation experience (i.e. before, at the same time as, or after obtaining sign-off on your proposal).

Such a device may be a welcome addition to your dissertation proposal as an appendix (Chapter 6) not only because it demonstrates your competence and the feasibility of your work but also it is emblematic of your willingness to be more transparent with your advisor/s. What is more, you can tailor such a table to the specificities of each method. For instance, Table 7.1 includes provision for 'archiving/securing'. If your research is qualitative in nature you may have recordings or other visual data (such as photographs or video) that you may have to retain and which you have assured your participants will be securely stored as your commitment to conducting research ethically (Chapter 9). Your institution may also require you to submit your primary data, perhaps as a potential check for plagiarism.

We have also included three other often-overlooked aspects. First, provision is included for insurance and other unexpected contingencies. Second, some students fail to check for data-entry errors or for inconsistencies in answers (perhaps associated with filter questions, see Chapter 12). Others fail to recognize the time required to prepare their data for presentation in the final dissertation text (Chapter 14). Raw output from some statistical packages may not be designed for direct input into your text. Qualitative research requires

Table 7.1 A framework for estimating the time required for data-related tasks

Task	Time (days)	Notes
Design		
Pilot		
Collection		
Coding		
Entry		
Quality checks		
Analysis/results		
Interpretation/findings		
Preparing for presentation		
Archiving/securing		
Insurance		
TOTAL		

Source: authors

you to think carefully about what quotations, exhibits or artefacts you may wish to employ from your raw data in order to support your analysis.

Table 7.2 presents a dissection of the time required for a hypothetical questionnaire survey of 400 respondents, while Table 7.3 offers a similar breakdown of a programme of 20 semi-structured interviews. Both numbers are deliberately chosen because students routinely underestimate the time required to collect and work with data, and as a result they often set themselves highly ambitious targets. Sometimes these are wholly unrealistic and we do not, for a moment, suggest that you should overdo it! Let's be clear, these are indicative examples, but they are helpful in benchmarking what *you* may be able to do in the time available for *your* undergraduate dissertation.

In both examples, the student would require seven full working weeks to design, conduct and write up the empirical work. To put this into context, the student would be working on a full-time basis for nearly two months (assuming a regular working pattern of the type we outlined above). This is without any other distractions or tasks related to the dissertation or the student's programme of studies. If the institution requires the student to spend 50 percent of time on the dissertation and 50 percent on other aspects of the degree programme, this work would take 14 weeks, or over a quarter of a year. This is quite a protracted period even if it were not interrupted by external factors such as illness, public holidays or even vacation.

A dissection of this type serves to shatter the myth that some students have which is that, in some way or another, qualitative research is a 'soft option'. In fact, qualitative research is arguably the harder option of the two. Consider Table 7.3: on average a half day has been devoted to each interview, the student assumes that each interview will take half an hour, and transcription will take three hours for every hour of taped interview. In

Table 7.2 The possible time required for questionnaire-based research

Task	Time (days)	Notes/assumptions
Design	1.5	Based on reading, consultation with methods notes and books
Pilot	1.5	Includes on-street trial, my assessment, discussion with advisor, adjustments, retest (?)
Collection	10	On-street from 08.30 to 17.30, 5 completed questionnaires every hour, 40/day
Coding	0.5	Code book and set up statistical package
Entry	8	Based on 50/day (might get quicker?)
Quality checks	0.5	Run simple frequencies and crosstabs to reveal errors and extent of missing values
Analysis/results	5	About a week of intensive analysis
Interpretation/findings	3	Time to write up parts of data analysis and conclusion. Comparison with existing studies
Preparing for presentation	1	Selecting which output to present and prepare for final dissertation text, in particular appendices
Archiving/securing	0.5	Half day spent on tidying up data, back-ups etc.
Insurance	3.5	c.10% of estimate
TOTAL	35	Seven full working weeks or equivalent!

Source: authors

the event that the average interview length were an hour and each hour of tape took five hours to transcribe, the student would have 20 hours of interviews and 100 hours of transcription! That is, based on transcription alone, the empirical work would be 70 hours, or nearly two working weeks longer before any additions have been made for analysis and interpretation. This makes the line for contingencies look a little mean. What is more, the estimates for analysis and results are precisely that: estimates. The actual time taken for these exercises may be much shorter (for instance, if clear themes emerge) or much longer (e.g. if there is considerable dissonance or if it's difficult to make sense of patterns in the data). Therefore, in some respects planning a programme of qualitative research involving interviews can be a more demanding and less predictable task than a questionnaire survey.

Of course, both are extreme examples, but they highlight the need to make sure that you understand the time you have at your disposal, that you monitor time carefully, and that you review the quality and quantity of data needed in order to address your objectives with your advisor/s. In the case of Table 7.3, if the student found that the early interviews

Table 7.3 The possible time required for a programme of semi-structured interviews

Task	Time (days)	Notes/assumptions
Design	1.5	Based on reading, consultation with methods notes and books
Pilot	1.5	Includes trial of interview, my assessment, discussion with advisor, adjustments, retest (?)
Collection	10	Half day for each interview roughly to include travel to interviewee, set up recording equipment, informed consent and preliminaries etc.
Entry	4	Based on tape:transcription ratio of 1:3. 30 minutes per interview = 600 minutes of taped interviews, or 10 hours = 30 hours transcription
Quality checks	1.5	No facility for entry (under transcription) but still need to check transcripts for accuracy
Analysis/results	10	Need to read and reread interviews, devise themes and codes, draw inferences from data. About 0.5 days/interview
Interpretation/findings	3	Time to write up parts of data analysis and conclusion. Comparison with existing studies
Archiving/securing	1.5	Need to ensure that informed consent forms, transcripts and tapes are deposited with programme director, as regulations require
Insurance	2	4 x 0.5 days as contingency for cancellations
TOTAL	35	Seven full working weeks or equivalent!

Source: authors

were taking twice as long as was budgeted, it would be important to reappraise the scope and feasibility of the planned data collection at the earliest appropriate point (let's say after five interviews). With the help of their advisor/s, an appropriately revised programme could be devised. In this context then, if you are interested in conducting qualitative research, ask yourself: how good are my listening comprehension skills? There are various estimates for the time taken to transcribe each hour of tape. We have seen figures ranging from one hour of taped dialogue requiring three hours of transcription (1:3) to one hour of taped dialogue taking six hours to transcribe (1:6), but this also depends on such factors as the type of accent, dialect, vocabulary and language being used, the clarity of recording (i.e. quality of microphone, interviewee's speech) and whether English is a native or second language.

As a final observation from Tables 7.2 and 7.3, you should note that if you halve the number of questionnaires or interviews, you do not necessarily halve the total time for data-related activity in your dissertation. Rather, any savings would be made in data collection and entry. For instance, the speed by which a sample of completed questionnaires

Table 7.4 The possible time required for six focus (i.e. discussion) groups

Task	Time (days)	Notes/assumptions
Design	4	Needs both a topic guide (to steer the discussion) and a recruitment questionnaire (to ensure appropriate panel composition). Pilot recruitment questionnaire and topic guide with peer group
Recruitment	4	Desire 6–8 participants per group. Recruit 60 potential participants (6 ×10) assuming 33% drop out between approach and groups. Recruit 15/day i.e. 2/hr
Preparation and planning	3	Scouting venues, comparing costs, booking venues, preparing visual aids, prompts, ordering F&B, arranging expenses and incentives
Collection	3	Two groups run per day, one in the afternoon (2.00–4.00 p.m.), one in the early evening (5.30–7.30 p.m.)
Entry	8	Each focus group to generate 2 hours of tape recording. As multiple voices on tape, assume that 5 hours transcription per hour of tape
Quality checks	1	No facility for transcription (under entry) but still need to check transcripts for accuracy
Analysis/results	3	Need to read and re-read transcripts, devise themes and codes, draw inferences from data. About 0.5 days/group
Interpretation/findings	3	Time to write up parts of data analysis and conclusion. Comparison with existing studies
Archiving/securing	1	Need to ensure that informed consent forms, transcripts and tapes are deposited with programme director, as regulations require
TOTAL	30	Six full working weeks or equivalent!

Source: authors

can be collected clearly increases notably if you are able to deploy more people: with two people, the time should halve; with three people, it takes a third of the time; and so on. This assumes that the research workers are appropriately deployed and that potential participants are not approached by more than one member of the research team. However, you will also need to invest some time in ensuring that the team is appropriately briefed and conducting the data collection to the same standards through a screening of the completed questionnaires and the entered 'raw data'. Focus (or discussion) groups are common suggestions among students to reflect their greater popularity in many forms of social research. Some of the appeal relates to the perceived efficiency gains of gathering a number of people at a venue at a given time. As Table 7.4 indicates, from a time-management perspective at least, the savings may not be as great as first thought. Moreover, these are put under even sharper focus when the financial costs are considered (Chapter 5) and some of the risks of no-shows from the recruitment exercise are considered.

To mix or not to mix? That is the question . . .

As you will have noticed, so far we have hedged our bets. We have not made a recommendation in favour of an exclusive (quantitative or qualitative) approach or the use of mixed methods. Many student dissertations are described as 'mixed methods' dissertations because they use some form of quantitative and qualitative data analysis to one degree or another in order to address their aim/s and objectives, as well as any research questions and hypotheses (see Chapter 5). As Barbour (2008: 151) poignantly observes, the mixing of methods is 'often employed in order to compensate for the perceived shortcomings of stand-alone methods, with the aim of either providing a more complete picture or enhancing coverage'.

To be precise, some of these dissertations may not take a 'mixed methods' approach, but rather be multi-methods studies: the distinction is that a mixed methods approach uses and integrates several methods as fully as possible while multi-methods use main and secondary methods with far lower, perhaps even minimal, integration. As Bryman (2007) argues, there is a wide variety of barriers to integrating mixed methods and insufficient attention has been devoted to writing up such work. Barbour (2008) demonstrates that there are many operational and intellectual challenges to be overcome if the methods and the data they yield are to be integrated efficiently and effectively. These include issues of how to reconcile different disciplinary traditions and other paradigms driving the use of methods (Feilzer 2009; Bergman 2011), the philosophy of science underpinning best partner mixed methods research (Johnson et al 2007), and the most appropriate strategy for blending methods to produce relevant knowledge (Johnson et al 2007). Many students forget that the sequence of deploying the methods, their relative importance in delivering objectives, and the capacity for one method and the experience of its application to inform the implementation of another, require careful tactical consideration.

There are therefore some important consequences of using more than one method for the planning of your time. The first is that it may be infeasible within the scope of your dissertation. This is the most common reason why students eschew this strategy in under-graduate dissertations. Furthermore, we have experienced several cases of students who have been determined to employ mixed methods but who have subsequently discovered they are trying to do too much, and they have had to modify (i.e. 'downscale') their objectives as a result.

You may be forgiven for thinking that mixed methods may present you with some significant economies of scale; that is, you will save time developing two or more survey instruments based on the same basic set of aim/s and objectives. However, in both mixed and multi-methods approaches you have to spend time designing, trialling and operating each of the methods properly. Likewise, you have to spend time coding, entering and analyzing the data they generate. A further investment of time is required to compare the results and findings from each of the methods, and this varies depending on the level of integration desired between the respective methods.

Time saving usually only occurs if financial resources constrain the size of the samples for each method employed. Data collection, coding and entry periods are likely to shorten, but the development, piloting, analysis and reporting times are likely to remain broadly the same as before. There is no way around it: however relevant to your objectives (and research questions) or intellectually attractive the use of more than one method may appear, this approach will place significant demands on your time budget (as well as your financial and human resources).

Latent time considerations

Beyond the obvious dimensions of data collection and analysis, there are several necessary but less obvious actions associated with your empirical research that can take more time than you anticipate, and they can sometimes introduce vital delays to your project:

- *Ethics approval* – while you can, and in most cases are required to, include an ethics statement in your research proposal (Chapter 6), it may only be scrutinized by your department, school or faculty once the basic proposal has been endorsed by your advisors. You may have to submit drafts (and revisions) of your survey instruments and sampling strategy and suspend further work on your pilot studies and full data collection until they are approved by the ethics officer or committee.
- *Criminal records checks* – if you are proposing to work with legal minors (i.e. children) or other vulnerable members of the community or in certain venues (e.g. schools, hospitals), you may be required to present certification from the authorities that you are a suitable person.
- *Health and safety approval* – again, such an assessment should be part of your dissertation proposal (Chapter 6). Depending on the insurance regulations at your institution you may be compelled to complete this assessment several times until your advisor and/or an expert panel are satisfied (or you run the risk of conducting your work uninsured).
- *Events, crises and disasters* – these can intervene during your study. In the best case, they may require you to suspend your data collection while the situation stabilizes and perhaps even returns to normal. In the worse case, they may go on much longer than anticipated, perhaps even throughout the entire period available to your dissertation. In which case, you may be required to change your topic altogether (Chapter 3) or significantly adjust your aim/s and objectives as well as methods (Chapters 5 and 6), perhaps requiring you to enter another dissertation proposal (Chapter 6). In this case, your advisor/s may suggest that your mitigating circumstances warrant an extension to your deadline (Chapter 10).
- *Getting access to dedicated software* – you may require specialist software packages (e.g. statistics, qualitative analysis or geographical information systems – GIS) that are only available on restricted machines by virtue of institutional subscriptions. As such, you may need to book time to use the relevant computing facilities. This may not necessarily delay your project but it will require tighter time management to

assume you have the necessary data collected and perhaps even processed by the time your appointed time comes around. Similarly, you will need to ensure that you book enough time to complete your work. Otherwise, you may find you have to wait for the next opportunity.

*   *Specialist training needs* – do you require any specific additional training throughout the course of your work? Are there new methods and techniques that are vital to your research but which you have not previously learned or experienced? Not only will training courses take time (perhaps a day or two), but also they may require you to extend the time you need to design and implement your survey instruments and possibly to conduct your analysis.
*   *Lost or damaged data* – you should always keep your data safe and secure, and back-up digital copies on a regular basis (Chapter 14). After all, think of the time and cost your data have taken to generate. Sometimes students lose questionnaires, Dictaphone tapes and memory storage devices on which their raw and processed data resides. This is irreplaceable and, without back-up, basically means you have to start data collection once more (see Box 14.1).

Scheduling your time

With the time required for each component in your dissertation estimated, the final step (for the drafting of your research proposal at least – see Chapter 6) is to map out what you are going to do and when.

In general, it is important to note that to date we have encouraged you to think in a task-based, almost cellular and linear manner about the time you need for each operation. This has been deliberate, as a means of provoking you to think about the overall feasibility of completing your work in the time you have available to your dissertation. However, in reality you may undertake several tasks at once when you are working on your dissertation. For instance, at the same time that you are drafting, developing and piloting your questionnaire survey, you may be finishing off your literature review as well as starting to think about the intricacies of the array of statistical methods and techniques you want to run.

This is a perfectly legitimate approach and it is one that should encourage continuity in your dissertation so that: the aim/s and objectives eventually articulated in your introduction are justified in the literature; they drive the choice of methods and their execution; and ultimately they shape the data analysis, results and findings. Several technologies exist to help you map out what you are going to do and when. At the most basic level, you may wish to enter an indicative timetable in tabular form with tasks and time set out (based on days or weeks).

More intricate approaches may also be employed for your proposal and to help you manage your research. Gantt Charts and Critical Path Method (CPM, Burke 2003: 131–53) are two common approaches in project management. They set out graphically the sequence of tasks, with their time requirements. Gantt Charts broadly plot time duration through a bar indexed by calendar position (on the y-axis) against the nature of tasks (on the x-axis). The Critical Path Method, breaks down a project into various packages of

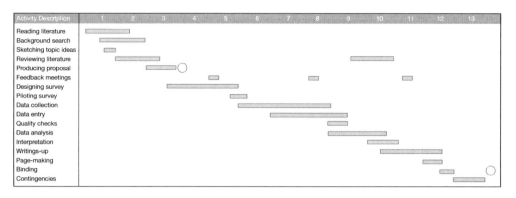

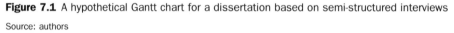

Figure 7.1 A hypothetical Gantt chart for a dissertation based on semi-structured interviews

Source: authors

activities (i.e. tasks) and maps out the paths and associated time required to achieve particular objectives in a 'network diagram' – basically a sophisticated flowchart.

Both technologies are helpful management tools for monitoring and evaluating the progress of your work towards completion of your dissertation. They are frequently used in planning research projects, albeit more commonly at PhD level because the size and scale of the undertaking is much greater. Notwithstanding, if you are familiar with either concept and you have access to proprietary software such as Microsoft Project, they are useful additions to your research proposal, although not compulsory elements by any means.

Figure 7.1 presents an example of a Gantt chart based on a dissertation informed by semi-structured interviews. In fact, the data within Table 7.3 have been programmed into the Gantt Chart which was produced in Microsoft Excel. The basic parameters are that the student had the final semester (13 weeks, or a quarter of a year) to complete the dissertation. A briefing meeting took place at the end of the previous semester where all students received their dissertation handbooks from the programme staff, but no advisors were appointed. Instead the student was asked to submit a proposal by the end of week three, and the dissertation at the end of week 13 (which was set aside as insurance). According to the handbook, meetings with an advisor should take place at the start of week five (to provide an opportunity for an advisor to be identified, assessment of the proposal, and feedback to be discussed), week eight (mainly to discuss data collection) and week 11 (to discuss analysis and the final steps). Other meetings could take place by agreement with the advisor so there was little point adding them.

When read with Table 7.3, you should note how the bars indicate the weeks during which certain tasks will take place; that is, this chart is indicative. For instance, within Table 7.3, three days were set aside for design and piloting but these were spread across weeks three to five by the student. This was a clever move: although some preliminary design work could be done to inform the proposal, there was no point completing the survey or piloting it until the advisor's feedback was obtained. Likewise, data collection (i.e. the interviews) and data entry (i.e. transcribing) were spread across three weeks so that, when not collecting data the student would be transcribing. Towards the end of the transcribing, the student could also begin reading through the transcripts as checks for quality assurance and as a re-familiarization ahead of coding (i.e. analysis) and interpretation. The latter would run in parallel to revisiting the literature review to contextualize the results as findings, and in parallel to writing up. It would be important to explain how the analytical approach worked while it was fresh in the memory. If the student could keep to the schedule, the timing of the last meeting would be helpful. It would offer a chance to discuss any last issues encountered in writing up and producing the final text.

A final thought: the complex relationship between time and successful dissertations

If you devote less time than the recommended total allowance, you may well be increasing the risk that you produce a poorer dissertation. All is not necessarily gloomy but this depends on you and your skill set. You may still pass – or even produce a highly successful dissertation – if you are efficient. You may have mastered the art of time management. What is more, this combination of efficient working with the delivery of effective results should hold you in good stead in the future.

Alternatively, if you are really keen to produce a highly successful dissertation, you may wish to invest more time than your programme recommends. There are, however, some significant issues for you to consider before you adopt this strategy. For instance, from where are you sourcing this additional time? What other activities have you had to sacrifice and hence what are the opportunity costs? These can literally be financial if you are forced to reduce the number of hours you work in paid employment. There can also be negative impacts on the rest of your academic work (or even your personal life). For instance, if you are conducting your dissertation in parallel to taught modules throughout your final year, you may be tempted to concentrate on your dissertation at the relative expense of studying effectively for your other modules. Will the results of your other (taught) modules suffer as a consequence? In turn, how may your final degree result be affected because you have prioritized your dissertation over your remaining modules?

In many programmes the dissertation has significant weighting in the final degree result because it carries more credits than other modules or courses. Understandably, as a result students are sometimes tempted to concentrate their energies more on delivering a successful dissertation. They perceive it as increasing their likelihood of attaining a higher degree result. There are dangers here, too. There is no simple linear relationship between time invested and the success of your dissertation. Just because you spend more time on your dissertation, your final result may not necessarily be proportionately much higher. Sadly, if you spend more than the recommended time there is no simple guarantee against your making mistakes of understanding or execution, however simple or indeed serious. Even were these not to present, the 'law of diminishing returns' is likely to come into operation.

The chapter at a glance

The main learning points of this chapter are that:

- Although time is a major consideration in the design and execution of your research, it is also a finite resource.
- You need to plan the use of your time wisely and fully, preferably before you submit your dissertation proposal.
- You should budget your time based on a breakdown of the various tasks involved in the dissertation, a realistic assessment of the time to complete each one successfully, and a consideration of their relative importance in the process.

Dissertation checklist

Before you go further in your work, check you:

1.	Found out how much time you are expected to devote to your dissertation in total and over what period.	
2.	Thought about how your dissertation work is going to fit with your other academic work and commitments.	
3.	Set aside an appropriate amount of time for each task.	
4.	Factored in time for unexpected contingencies and hidden delays.	
5.	Have not underestimated the time it will take to complete your data collection and data analysis.	

8

YOUR PERSONAL SAFETY

Learning outcomes

By the end of this chapter you will be able to:

- Identify hazards and associated probabilities affecting your research.
- Conduct a basic risk assessment of your proposed programme of work.
- Compile and maintain a simple risk register.
- Understand some of your main responsibilities as a 'research citizen'.

Keep calm and carry on

Don't panic! Each year literally hundreds of dissertations in tourism around the world are conducted without a problem for individual researchers. Almost inevitably, though, there are some students who encounter difficulties in the course of conducting their research. We cannot put a figure on how many because there has simply been no comprehensive research on the subject. Suffice to say, from our experience the proportion tends to be very low, and we are of the opinion that this proportion can – and indeed should – be minimized if students (with their advisors) take the correct measures to identify and manage the risks involved in conducting their research.

Forewarned is very much forearmed, and it is those students who prepare thoroughly for their research and the contingencies this may present, that tend to be those who are less likely to be affected by problems. Your dissertation concerns your personal safety, and it is quite easy to be blasé about such matters. After all, what were the probabilities of a volcanic eruption in Iceland or a tsunami in Japan, moreover one that resulted in a major nuclear incident (see Box 7.1)? It is not uncommon to find students who don't want to be bothered thinking about such matters because 'it will never happen to them' or because it is so desperately 'uncool' (Figure 8.1). Sadly, those who are least prepared are often the most vulnerable to the fullest extent of unexpected occurrences.

The purpose of this chapter is to urge you to take your personal safety seriously during your dissertation research. In one respect, the chapter takes a reasonably predictable

Figure 8.1 Forget the possibility of crocodiles, enjoy the view down at the riverside!
Source: Tim Coles

position: it encourages you to consider the risks to you and your work properly during the planning *and* execution of your independent research. Practising good 'research citizenship' is important, as we'll discover in the next three chapters.

Finally by way of introduction, you should consider what follows in this chapter as an introduction to your personal safety. It is not, nor is it intended to be, an extensive or comprehensive treatment of these issues, or those surrounding fieldwork. Moreover, this chapter is not intended to replace procedures locally at your institution, but rather its intention is simply just to raise your awareness of issues which sadly so many students take for granted.

Risks at home and away

There are many reasons why students are blasé about their personal safety. Sometimes this is sheer bravado; other times it is based on a heady cocktail of experience of travel (perhaps during a 'gap year'), or prior knowledge of the locations where they're going to conduct their research.

Events happen and they happen at home just like they happen overseas; they happen in familiar places not just alien environments (see Box 8.1). They happen in the ordinary closed, confined spaces of offices and libraries and they happen outside, as it were 'in the field'. This rather idiomatic English expression, 'fieldwork', is inherited from the physical sciences and in particular from geography, to denote scholarly work away from

Box 8.1 Safety begins at home

Exeter is one of the safer cities in the UK and its attraction to students is its high quality of life and standard of living. In principle this makes it an ideal place to conduct a survey.

One of our students wanted to gauge local residents' views of the opening of an attraction in the city. A representative sample was desired. Initial thoughts about the sampling strategy were to find an appropriate location in the city centre at which to complete a series of questionnaires on-street, face-to-face and with randomly-chosen passers-by.

Problems of time, the length of the questionnaire, and doubts about whether some social groups visit the city centre as much as others led to the replacement of the on-street strategy with 'drop-off and collect'; that is, to post the questionnaires and cover letter through the letter box and return in person a couple of days later at an appointed time to collect the completed questionnaires from the doorstep. This was deemed feasible because the city has a population of around 120,000 people spread across about 20 square miles.

With 400 questionnaires to distribute, the student first calculated how many were needed for each ward to stratify the sample; thereafter, addresses within each ward were selected at random. Ready to start dropping off the questionnaires, the student was somewhat taken aback when asked by the advisors to explain what safety measures were being taken to manage the research. After all, by this point the student had lived in Exeter for nearly three years and knew it reasonably well. What could possibly happen?

Exeter may be a low-crime city but it is not a zero-crime city, it was explained. A strict policy was agreed that the student would not enter a home, however welcoming or well-intentioned the invitation was meant. The advisors were given the student's mobile phone number. For information, the student provided the advisors with a list of addresses and dates on which they would be visited in the morning or afternoon to drop off the questionnaires. A friend in the same dormitory was also informed by the student and asked to call the advisors if the student did not return with an hour of the scheduled return time.

Pick-up was to take place in the early evening in order to boost response rates (i.e. after people had returned from work) but it was early spring and it becomes dark quite early. Once more a list of addresses and dates was provided but this time the student asked another friend to help in the collection process with a view to speeding it up and assuring each other's safety. Suffice to say nothing went awry but better safe than sorry.

the confines of the classroom, office or laboratory studying phenomena *in situ*, in their natural habitats (see also Hall 2011a).

In the social sciences, studying tourism subjects *in situ* may involve interacting with hosts, guests and/or communities in rural environments such as on farms, at lakes and on river banks, within mountain ranges, remote deserts or on coastlines. It may also involve interactions in urban environments, in shopping malls, museums and visitors centres, convention centres and exhibitions, concert halls, sports venues and in airports. Each presents its own distinctive forms of risk to the prospective researcher. Indeed, an interesting but frequently overlooked feature of recent work on crisis management within tourism is just how frequently and regularly events impact on the tourism sector around the world (Sönmez 1998; Coles 2004). Catastrophic floods, hurricanes and other extreme weather episodes, earthquakes and tsunami, wars and conflicts, not to say infectious diseases and other biohazards are scarcely off the media radar these days (Ritchie 2009; Hall 2010; 2011b).

Risk assessment and risk management

It does not matter whether you are conducting a piece of fieldwork on the future of the British coastal resort in the Kent town of Margate, archival work in Library of Congress, Washington DC on the history of outbound tour operators from the United States, or the willingness to pay among ecotourists in the Taman Negara national park in Malaysia, there are hazards in all types of research and the environments in which it takes place. What changes, of course, is the precise array of hazards, the exact probability that you will encounter them, and the severity of the impacts should they occur.

This combination of hazard, probability and severity is the basis for standard approaches to risk assessment and, perhaps more importantly, its sometimes unfairly overlooked companion concept, risk management. It is no use predicting that something might well happen without having the measures in place to respond if (or indeed when) it does!

You may not be surprised to learn that there are many approaches to risk assessment and management requiring varying complexity and detail. First of all, you should check with your institution your obligations to consider risk and your personal safety as well as at what times (i.e. during the planning and/or continuously updated throughout the research).

For the compilation of your research proposal, it is highly likely that you will have to complete some form of 'risk register' (Table 8.1); that is, a table comprising the following fields with a commentary:

- Name of hazard.
- Probability of hazard.
- Severity of hazard.
- Risk Score.
- Level of risk.
- Risk-management measures.

Table 8.1 An example of a risk register

Name	Street crime	Crimes against the person	etc. . . .
Probability	1	1	etc. . . .
Severity	2	3	etc. . . .
Risk score	2	3	etc. . . .
Risk level	Low	Medium	etc. . . .
Management	General vigilance Avoiding known hotspots Planning routes carefully	• Not to enter any home • Mobile number to advisors • Addresses/visit dates to advisors • Friend to call if not back within hour of planned return • Second friend to accompany evening collection	etc. . . .
Comments	Low-crime city	Very low probability but potentially massive impact on me personally. Not taking it too lightly	etc. . . .

Probability scores: 1 = low, 2 = medium, 3 = high
Severity scores: 1 = low, 2 = medium, 3 = high
Risk levels: 1–2 = low, 3–4 medium, 6 = high, 9 = extreme

Source: authors (adapted from Burke 2003)

In starting to complete your risk register, three things are initially important. First, you should think through all of the types of hazards that you think you may potentially encounter while you are conducting your research. As with other aspects of research planning, be comprehensive. Think widely about what might occur. This is not being paranoid, it is being sensible.

It is unlikely that the data will be available, or indeed that you would have the time or inclination to research the precise statistical probabilities that those hazards will befall you during the period of your research. Second, therefore, in many a risk assessment, broad categories are attributed to the probability of a particular hazard occurring: they are given a basic numerical score, and this is entered as a second piece of data in each line of the register (see Burke 2003). Some hazards are more dangerous and hence have a greater potential impact on you (and your research), than others. Third, then, the severity of each hazard can be broadly graded, given a numerical score, and entered in your register.

Calculating risk

In most systems, the value for the severity of the hazard is then multiplied by the value for the probability that the hazard may occur, in order to compute an overall score for each hazard. Once this has been calculated, it is compared to a series of preordained thresholds (class intervals) that suggest how significant each risk is to you and your work (or not as the case may be) and hence the need for any further action.

Risk managment measures

The nature and level of response has to be commensurate with the level of risk you have identified. In some cases, risk has to mitigated; that is, a certain level of risk is accepted and behaviour is changed to ensure that the risk does not result in negative outcomes for you or your work. For instance, there may be an above-average probability of your catching malaria in your fieldwork location and you cannot avoid being bitten by mosquitoes. Nevertheless, when you travel you should ensure you take counter-measures such as covering-up, the use of insect repellents, and the right prophylaxis for you!

In other circumstances, risk can be reduced (or virtually eliminated) by adapting your behaviour to better suit the circumstances you are likely to encounter. If your fieldwork location is flood-prone, you might want to adjust the timing of your visit if this is at all possible. Hazards and levels of risk vary throughout the year and over time. For example, equatorial regions experience their highest temperatures and greatest rainfall around the vernal equinoxes (March to April and September to October). In Kenya and Tanzania these are sometimes referred to as the 'little rains' and 'big rains'. Standard guidebook prescriptions do not suggest that it is the wrong time to visit, but rather that there are restrictions to travel during these periods and hence it is more productive to visit outside the rainy season.

Table 8.1 shows an example of a risk register based on Box 8.1 and the student's work on a questionnaire survey using 'drop-off and collect'. This just shows two types of risk and, in reality, the risk register would be longer than the fragment provided here. It is important to note the honesty of the student as well as the somewhat crude nature of the device. For example, we all know why students should not enter homes and the potential dangers that this puts them under personally; the consequences to the individual could be horrendous. The probability of this happening is thankfully very low, much lower than street crime, yet because of the somewhat unrefined scoring system it is given the same probability rating. More sophisticated scoring systems exist and they may be used by your institution but we should not deviate from the main point which is to be honest and thorough in your reflections of what just might happen and how you should deal with it.

Rights and responsibilities

In the main, risk assessment and management in research is about reflexivity, but it is also about good 'research citizenship'. Rights are accompanied by responsibilities in many standard conceptualizations of modern citizenship (Coles 2008).

Bestowed by those who designed and operate your programme of studies, you have the right to conduct a programme of independent research supported by advice from members of staff, access to facilities such as IT, printing and photocopying, library resources, and the cover of your research work under institutional insurance policies (as well as your own).

Figure 8.2 There may be a reason why the site is roped off?

Source: Tim Coles

These rights are accompanied by a series of responsibilities that are expected of you. You have responsibility for your own well-being and for those around you like your research subjects (i.e. those who complete your questionnaires or agree to be interviewed or participate in focus groups). You have a responsibility to your advisor/s not only to conduct your research to the highest academic (Chapter 10) but also ethical standards (Chapter 9). You also have a responsibility to any sponsors who are funding your research, as well as to your friends and family or whoever else is supporting your endeavours.

With respect to risk assessment and management, responsibility starts with thoughtfulness and identifying the full array of hazards you may encounter. You have to be diligent and conscientious in order to document them and to think through the likelihood that the hazards may affect you and/or those around you. You need to take the completion of your 'risk assessment' seriously and ensure it is submitted in a full and timely manner for scrutiny. After all, you expect those assessing your work to demonstrate the highest standards of professional practice in reviewing your risk assessment and your wider research proposal. As such, they have the right to expect that you will show them the same courtesy by taking this exercise seriously.

Your responsibilities do not end with the formal sign-off by your advisor/s and/or a health and safety panel. You have a responsibility to act safely whenever and wherever you are conducting your dissertation work. As Figure 8.2 would suggest, there is no use identifying the potential risks associated with conducting fieldwork in the Southern Alps of New Zealand, if you flagrantly ignore the official advice given by the local authorities.

As a good 'research citizen' you should continue to monitor the hazards, probabilities and levels of risk as they relate to your work. You should continually update your 'risk register' as your research progresses. In some cases. this is necessary when you are 'in the field'. The hazards and risks you encounter when you come to conduct your fieldwork may not be entirely the same as when you submitted your original proposal. For example, the rains may come earlier than expected, a volcano may unexpectedly erupt or an earthquake may happen for the first time in a century. As a result, and where necessary, you may need to introduce new or revised approaches to managing emergent and changing risks, and you may need to inform your advisor/s of these changes (especially when you are away from your higher education institution).

Assessment of risk assessments

In most dissertation programmes your risk assessment will be part of the package of papers that comprise the research proposal (see Chapter 6). In some cases, you may be asked to submit your risk assessment independently from the research proposal in a parallel process of assessment. Whatever the precise mechanism, there are at least two important layers of scrutiny:

- First, and perhaps most importantly, there is *your* assessment because *you* should have spent significant time thinking through the issues, and *you* should know the intricacies of *your* research project better than anyone else.
- The second layer comprises external assessment of risk by your advisor/s (Chapter 10) and/or by an external expert officer or panel (i.e. departmental, faculty or even university) on health and safety.

If your advisor/s and/or other external assessors are not satisfied, they may refer your document back to you for revision and resubmission before they are prepared to sign-off the further progression of your work. This may introduce a vital delay to your project (see Chapter 7) while you undertake remedial work and await the next round of assessment. The delay may also have the 'knock on' consequence of deferring the approval of any travel insurance cover offered by the institution centrally, and hence your departure date.

Thus, it is an extremely unwise strategy to enter a very provisional document with the expectation that your advisor/s or other parties may complete the work for you. Some students we have supervised have been under the sad misapprehension that they will receive detailed listings on what they may have overlooked; hence there is no need to perfect their submitted draft. Let's be clear: it is not the responsibility of your advisor/s and/or health and safety officers to think for you!

Overseas travel

Many students choose to travel overseas to conduct their fieldwork. After all, as we have noted in Chapter 1, for many students their dissertation represents a fantastic and perhaps once-in-a-lifetime opportunity to study the topic of their choosing. Some students embrace this by looking at subjects and destinations that they may not ordinarily include in their own private travel and tourism planning. The dissertation effectively becomes a form of business travel, with many if not all the usual things to consider in planning and preparing such a trip.

In many respects, buying an up-to-date travel guidebook from a major publisher will give you detailed and specific information about the destinations where you propose to conduct your research. Some general reminders are presented in what follows.

Travel documentation

You must make sure that you have an up-to-date passport and visas for entry to the countries in which you propose to conduct your research. In many cases, visas need to be obtained in good time before your departure. Not only are delays costly in terms of rescheduling travel and accommodation, but they also eat into your limited time window for completing your dissertation (Chapter 7). There is no guarantee that you will be offered an extension because you could not organize your travel arrangements properly. Conceivably, this may offer you an unfair advantage over a fellow student who decides to stay at home to conduct their research.

In terms of compiling and updating your risk register, one of the first actions you can take is to examine the foreign travel advisory service of your government before you consider travelling and, again, just before you travel. The United States' Travel.State.gov site provides this advice. In the case of the United Kingdom, the Foreign and Commonwealth Office (FCO, effectively the foreign ministry) provides information for UK citizens about travelling abroad. This includes a range of subjects, from general advice on safety from disease, to the threats of terrorism, to advice on travelling overseas to attend major events and sporting spectacles such as the soccer World Cup and Olympic Games which might be the subject of your research.

For UK citizens (i.e. passport holders), such a site is immensely important in so far as some insurance companies use the government's assessment of risk as their *de facto* positions (Coles 2008). If the FCO declares it unsafe to travel to a destination (whatever the merits of its decision), the insurance companies are likely to follow suit; thus, if you travel – even with an otherwise apparently valid policy – you may well be travelling without the cover you anticipated.

The foreign ministry service in your home country, as well as that in the destination country, is extremely useful in a third and final respect: immigration and entry. You should be careful to evaluate your status properly before you travel. Some countries (like Australia and the United States) require visitors to register their intention to travel (so they may be prescreened). Remember, not all short-term, temporary entry permits and schemes may allow you to conduct research, and you may be required to apply for a full visa. To be eligible to enter your proposed destination under the entry permit scheme, you may be required to demonstrate you have the (financial) means to support yourself as well as a return ticket home. Many entry permit schemes are assessed on arrival and you should check to ensure that your entry permit will allow you all the time and freedom of travel you require in order to complete your research.

Medical preparations

Ensuring that you are properly prepared medically is of vital importance and it is one immediate means by which you may mitigate medical risk before you leave. You have a responsibility to your institution and to insurers – not just to yourself, your family and friends – to ensure that you have received all the necessary inoculations, and that you have started courses of medication as directed by your physician in advance of your trip. If you fail to do so and/or if you travel against the direction of your physician you may find yourself facing a large medical bill if the insurance company declines to honour your claim in the event that something happens to you. Your physician and medical practice will be able to offer you advice on pre- and during-trip medication. There are several online facilities and advisory services, and generic advice is offered in most of the established travel guides. In fact, travel medicine is such an important issue that guidebooks are now being published solely devoted to travel health (Jones 2004).

Intercultural communications

As a final thought, you should consider the cultural context of the communities in which you are about to travel and conduct research. This may sound axiomatic because there is such a well-established body of knowledge dealing with host–guest encounters (Shaw and Williams 2002; 2004). This academic background knowledge is, however, no substitute for understanding more about your host society. This can include such issues as whether it is permissible to take photographs of buildings and people, the sort of behaviour which is acceptable on the streets, and some basic words to communicate more effectively with the people around you.

As standard, some basic advice of this nature is included in many, if not all, of the major and widely available tourist guidebooks. Nevertheless, as Scherle and Coles (2008) have shown, there is more to intercultural communications than the basics of meeting-and-greeting, on-street behaviour and photography (see also Reisinger and Turner 2003; Jack and Phipps 2005). They point out that there are different cultural conventions and expectations for professional and commercial relationships on either side of an intercultural dialogue. The key point is that, as a researcher, you have to be aware that different cultural conventions and expectations also mediate the research encounter. Moreover, the extent to which you are able to engage in cross-cultural dialogue may go a long way towards enriching your research. Not only will you understand more clearly why certain things happen, but you may also be welcomed more enthusiastically and hence given comparatively wider access to people or information (which enhances the chances that you will obtain the necessary information you need to make your work a success).

The chapter at a glance

The main learning points of this chapter are that:

- You should take your personal safety seriously when conducting your dissertation research.
- There are risks associated with all aspects of your research and, while they may be (comparatively) greater when you are conducting 'fieldwork', they are not restricted to this phase of your work.
- If you do not understand what is required of you in this regard, you should discuss matters with your advisor/s at the earliest opportunity.

Dissertation checklist

Before you go further in your work, check you have:

1.	Understood your safety responsibilities to your institution, your advisor/s and other stakeholders in your research.	
2.	Documented the full array of potential hazards that may affect your work.	
3.	Considered the probability that each hazard may occur.	
4.	Put in place measures to mitigate or avoid unacceptable risks.	
5.	Compiled a basic risk register.	
6.	Incorporated your personal safety in your research proposal.	
7.	Made yourself aware of how your health and safety declaration will be assessed.	

9

ETHICAL CONSIDERATIONS

Learning outcomes

By the end of this chapter you will be able to:

- Explain why a consideration of ethical issues is a vital aspect of independent research.
- Identify ethical issues associated with your dissertation.
- Address the demands of an ethical assessment exercise.
- Articulate more of your main responsibilities as a 'research citizen'.

Introduction

The purpose of this chapter is to lay a foundation for your understanding of ethical issues in tourism-related research as they relate to your dissertation. Conducting your research in an ethical manner is about ensuring that work is conducted to the highest standards of integrity and with the greatest respect for the various stakeholders involved in your research, including your subjects (i.e. the participants in your research), your advisor/s, other students, and the wider institution at which you are working. Acting in an ethical manner in research is about the general principles of what ought to be done whereas morals are about what it is right to do (Hart 2005). Through the structure of your degree, you have the right to conduct independent research but in return you must take your responsibilities seriously.

Ethics is not just a consideration for master's- and doctoral-level students. It is a key concern for students at all levels, especially undergraduates undertaking what might be the first of many steps in independent research. This chapter discusses why ethics are important to tourism research in general and how they factor in to your research in particular. Simply put, ethical considerations form an essential part of research governance in higher education and most, if not all, research requires some form of ethics clearance, yours included. One obvious opening move is to make yourself aware of the ethical regulations governing your dissertation or codes of conduct for independent research at your institution. This is important not least with respect to what Hart (2005: 163) terms the 'ethics of using the literature' which we encountered in Chapter 4. Subject associations and disciplinary groups also produce ethical codes of conduct which may assist you (Tarling 2006).

Ethical scrutiny in research

The reasons for ethical clearance are primarily threefold, namely:

- To ensure the research is conducted in an appropriate manner (usually defined through policy), and which is sensitive to the interests of participants, respondents and stakeholders (involved or otherwise).
- To ensure that the higher education institution (and your home department, school or centre) is aware of the scope of your research, which can allow it to help you to find your way through any potential issues that may arise as a result of your proposed approach.
- To confirm that data acquired empirically is collected in a reasonable manner (again, defined through the policy at your institution), and that the institution is generally aware that data are being collected and subsequently used.

Thus, ethics in research serve multiple purposes for the various stakeholders involved in your research. First, ethics and codes of conduct inform you of the appropriate boundaries to your enquiry. Second, they allow the subjects of your research to understand their rights and risks if or indeed when they decide to act as participants. Third, they enable institutions to properly monitor the types of research being produced by their students (and staff).

The researcher's boundary

Researchers have boundaries set around them by numerous third parties. For example, research involving children may require special ethical considerations that are time consuming and thus in need of careful planning to ensure appropriate conduct. Cultural sensitivities also play a role. In New Zealand, for example, proposed research involving Maori may (depending on the institution) need to be presented to appropriate groups or organizations. Some higher education institutions forbid interviewing (their own) local students unless special permission is sought.

Many of the policies concerning ethics and boundaries in independent research surround the use of participants as research subjects. In other words, these are the people to whom you may be distributing questions or interviewing at, say, the train station. There are often very strict guidelines that govern how you actually go about building and using a sample of participants in your research (Chapter 11). Thus, it is wise to establish these parameters as you develop your objectives and methodological approach (Chapters 3 and 6). Three of the more common restrictions relate to:

- *The age of participants* – there is variability in the age of matriculation country-by-country, and anything below this will usually require parental consent for participation. You may also be expected to have appropriate police clearance. In this instance, you should perhaps ask yourself, in the context of your initial research design (Chapter 3), why you would want to talk to anyone who was not an adult? However, there are clearly some cases in which legal minors may be appropriate to your topic.

- *The connecting of participants with the data they provide* – as a general rule of thumb, individuals should not be identifiable by name on a questionnaire or interview transcript. For practical purposes, this can be difficult to achieve. For example, sending out a questionnaire with a small portion at the end for the respondent to record their name and contact details (perhaps for the purposes of an incentive draw, for instance) automatically links names to responses. To get around this, simply cut off the section of the questionnaire that has this information so that the link between subject and data is immediately severed.

    For qualitative research, things can get a bit trickier. Standard ethical practice is to separate proper names from interview data so the interviewee's proper (i.e. real) name should not be attached to a transcript. Hence, it is necessary to develop of list of pseudonyms and proper names on a separate sheet of paper which is then kept in a secure, off-site (i.e. away from your primary data) location. It may be the case that you wish to conduct a follow-up interview with a respondent in order to clarify some responses. Doing so, however, means that you need to connect any new interview data from a second interview to the first interview with the same respondent. In order to do this, you would have to look up the pseudonym before appending the new material to the old. Once the research is finished, this master list would be destroyed.

- *Revealing of participants within the text through the data they provide* – protecting identities in raw data is one thing, but it is quite another when it comes time to writing up your research. As a general rule, participants should not be identifiable through what you write. Generally, they should be assured that you will make every effort that their identity will not be revealed. This is often stated in an information sheet or consent letter (discussed later) that they may be given and asked to sign. For both parties, this establishes the understanding that their anonymity will not be compromised and confidentiality will be maintained. To violate this by offering identifying characteristics (whether intentional or not) in your text is poor practice and could – in certain circumstances – even expose either you or your institution to legal proceedings. You may be tempted to ask why anonymity should even be preserved? Apart from basic institutional policy and legal reasons, you may find that respondents are less likely to participate if they are not assured of their anonymity.

    While maintaining anonymity is reasonably straightforward in principle, in practice it can often be quite a lot trickier, as Box 9.1 reveals. A proper consideration of ethics can leave with you with some awkward dilemmas to ponder carefully. For example, suppose a student is conducting research for her dissertation that involves discussing attitudes toward event planning in a small resort town. To maximize coverage of the issues, her

Box 9.1 Finding your way through the 'moral maze'

Our students like to travel. Many of them view their dissertation research as an opportunity to satisfy their wanderlust while conducting their empirical work. Just to explain to you how tricky ethical considerations can be and the importance of regularly revisiting your ethical position while you're conducting your research, let's introduce you to one of our students. Let's call this person, X, which is not an initial used in the person's name. X was interested in conducting research in an African country with an emergent tourism industry, with a view to working for a non-governmental organization (NGO) after graduation. A programme of semi-structured interviews and short questionnaires was designed and it received ethical approval.

The research aimed to examine how senior industry figures in the main destination understood a recent government strategy and blueprint for tourism development. Briefly put, X intended to examine the policy efficacy of a top-down approach. When it came to conducting the research, all of the interviewees had received a letter describing the intention of the research; they freely signed the consent forms with their guarantees of anonymity; and they were happy to be taped on the understanding that transcripts and tapes were locked away at first, then subsequently destroyed. None of this was exceptional.

What X found – but had not anticipated – was the sheer strength of opinion, in particular from expatriate managers, which was highly critical of the current government and its reliance on international consultants whom, it was alleged, simply did not understand what was happening 'on the ground' in the country. In X's view, several had used the cloak of anonymity to express views they would not have otherwise articulated publicly. This was highly interesting if not relevant to the research.

However, when it came to the analysis and writing up, X was presented with two connected dilemmas: some interviewees, including a member of the ministry, had asked to receive a copy of the dissertation as a condition of their participation; and because the community of practice (and hence sample size) was small and the comments were so specific, there was an extremely high chance that particular respondents could be recognized from their comments. To overcome this difficulty, reported speech had to be used, almost exclusively, and the text had to be screened carefully. With that problem solved, student X then worried that, in the absence of direct speech, the examiners may think it was 'made up'.

It all turned out for the best. For understandable reasons, the external examiner asked to see the transcripts and, after discussion, X's approach was commended. More importantly, as a result X was offered positions by several participants, including an NGO which had received the dissertation. This was because of the local sensitivities and understanding X had demonstrated in conducting the work.

Source: authors

sample of interviewees contains representatives from a variety of sectors, including accommodation, rental cars, airlines and local councils. Time and budgetary constraints (see Chapter 7) limit the total number of interviews she can do to perhaps one or two of each. Letters of consent have been signed by both parties ensuring anonymity. When it comes time to write up the results of her research, she finds that the opinions of the two accommodation providers are quite critical of the local council (for whatever reason) and would contribute greatly to her analysis. In this case, it is vital for her not to name the individuals whom she interviewed, but this may not be enough. If she omits the name, she may be tempted to identify the name of the establishment at which the individual works. Doing so, however, may easily compromise the individual, either publicly or internally within that establishment. Thus, the best course of action in this case is to omit the name of the establishment as well.

Put another way, consider this: if ever the name of an individual or their place of employment is absolutely critical to the final text of your dissertation, then special permission must be sought to use their names (following your institution's guidelines and policies, of course).

More often than not, however, proper names are irrelevant to the data produced and interpretations gleaned. For this reason, many researchers use pseudonyms to hide the name of the person or the organization with which they are connected. However, there are two other common means by which even 'anonymized' respondents might be identified:

1. Some researchers contextualize their analysis and the comments and positions they report by providing short 'thumbnail' sketches of their interviewees based on certain relevant socio-demographic characteristics. Sometimes these background variables are provided in tabulated form. Care must be taken in so far as the combination of particular attributes may be unique to an individual and hence create the possibility that their anonymity may be compromised.

2. As the previous example attested, in some cases tourism research is conducted in small communities of practice where many of the respondents know one another. What is more, they may be known for the particularity of their (perhaps divergent, radical or even downright outrageous) views. In which case, the use of 'verbatims' (i.e. direct quotes) may identify them. More subtle still, some respondents may employ particular vocabulary, phrases, idioms or other constructions that identify them to others. In this case, the use of reported speech or paraphrasing may be the most appropriate solution.

A cautionary note about secondary data: ethics in the blogosphere

We are often asked about the ethical considerations in using secondary sources (e.g., books, manuscripts, blogs, websites, etc.). This can be a grey area. On the one hand, the use of books, reports and journal articles in a dissertation is entirely ethical and it is made so by appropriate referencing and citation to avoid accusations of plagiarism or misrepresentation. On the other, the use of online material (e.g., blogs, discussion forums, websites) raises interesting issues. You might think that because this is in the public domain in open access, it is 'fair game', although it is not as simple as you may think.

For example, Duval (2008) utilized data from an online discussion forum in a discussion of how travellers portrayed themselves as Canadians. Naturally, some quotes were planned for inclusion in the text: in other words, the idea was to allow the 'respondents' (i.e. the forum posters) to use their own voices. The intention was, then, to individually contact each forum 'poster' whose words were desired and ask their permission, under the premise that, if they wrote it, they 'own' the words. A close reading of the forum's terms and conditions, however, revealed that, upon signing up, posters agreed to transfer ownership of the content to the owner of the forum. Thus, only permission from the forum owner to utilize what was said was required, sought and ultimately secured. The lesson here is that it is useful to investigate all avenues of obligation and rights of ownership, especially when it comes to online data.

The use of blog data is fraught with further ethical dilemmas. For instance, a blog owner may assert copyright over their work, despite the fact that it is publicly available. In other cases, they may assert what is known as 'creative commons' copyright (www.creativecommons.org), which comes with varying levels of circumstances for which material on the site can and cannot be used. It is entirely possible that your institution's ethics committee has not established a clear policy on these issues. In this case, it is your responsibility to acquire permission, where possible and where appropriate. Indeed, a rule of thumb that should be pervasive in any research is that you should seek 'informed consent': the more people who know about what it is you are researching (and in detail), the better.

The ethical rights of a participant/respondent

Respondents or participants are often given a set of 'rights' or expectations that they can demand of the researcher. Once again, this depends on individual institutional policy, but for our purposes we shall illustrate examples of where a respondent is afforded certain rights. These are usually outlined in a covering letter or consent form which is signed:

1. Respondents can withdraw from participating in the research at any time, without fear of retribution or harm.
2. Respondents' participation is usually voluntary.
3. The data they provide (in the form of, for instance, an interview) will be destroyed after a certain amount of time has elapsed.
4. Their anonymity will be preserved, if not totally then to the greatest extent possible.

There are others (e.g., remuneration agreements, if applicable) but these can be seen generally as the four key 'rights' afforded to respondents if they assist you as subjects by participating in your research.

Oversight: the institutional perspective

Higher education institutions have a great interest in the research that is conducted under their name. While they are clearly interested in their reputations, they have obligations to society at large and to the academic community specifically to ensure that all research conducted is fair, appropriate and conforms to established standards of ethical scrutiny. As such, systems are put in place to confirm that the research being conducted by their students (and staff) conforms to their stated policy. In turn, this is often informed by a combination of positional statements and best practices published by research councils, disciplinary associations or subject groups.

The manner by which ethical clearances are considered and awarded varies by institution. In some instances, individual departments or schools may be authorized to award ethical clearance, although higher-level institution-wide ethics committees may review and ratify such decisions as appropriate. In other cases, ethical clearance is exclusively granted by a central institutional-level committee. Importantly, ethical guidelines and policies should be available at your institution and you must follow these in your dissertation research. Your dissertation handbook should point you in the direction of these guidelines or even repeat them out for you.

If you require guidance on ethics, in addition to your advisor/s (Chapter 10) you may find that department- or school-level ethics officers may be available for consultation and assistance. As with issues of personal safety (Chapter 8), think for yourself first and prepare specific points for discussion based on your research before you meet. Do not expect your advisor/s or the ethics officer to come up with the answers for you!

You should be aware that applying for ethical clearance may take some time, largely because committees may only meet infrequently (e.g. once a month). Thus, you should factor in time for clearance in your project planning (Chapter 7). Technically you may not be allowed to begin your (empirical) research unless ethical approval is in place. Delays in clearance – perhaps because of an incomplete case or the need for clarification – can cost you important time.

Practicalities: putting together the ethics application

Obtaining ethical clearance will vary from country to country and institutionally; however, students are routinely required to demonstrate an acceptable level of ethical consideration in their research proposals (Chapter 6) and/or complete a dedicated ethical clearance form. Where both are requested, the former usually acts as a summary of the main issues, while the latter acts as a much more detailed exposition of the issues you are likely to encounter as well as the approaches and measures you will use to address them. Nevertheless, whatever the precise format or system your institution employs, there are some common elements which can be discussed in detail.

- *Description of the project* Without question, you will be asked to provide some basic details on the project or research you are undertaking. This is critical because it gives those responsible for assessing your consideration of ethical issues a sense of what you are trying to accomplish in your research and how you are proposing to do it. Ethical issues are context-specific and linked. They are revealed in aim/s, objectives and research questions (if you are using them), data-collection methods and analytical techniques, and how you intend to report your results and findings.

• *Ethical issues* Next it is common for ethics application forms to ask the researcher (sometimes called the 'principal investigator') to list, in detail, any ethical issues which s/he believes will arise during the research and how they are going to be addressed. The list can sometimes be quite lengthy, but it is important that you list every possible ethical eventuality. One way to do this is to consider the likely ethical issues across the entire dissertation lifecycle; that is, consider the ethical issues for each of the main components of your dissertation and the attendant tasks (Chapter 2).

Not surprisingly, this is where you might seek some assistance before submitting the application. Some (but certainly not all) of the issues that you may need to consider are:

• Where and how (securely) will your data be stored? Who will have access to it?

• What are the risks to the people or organizations participating in your research?

• Will any qualitative interviews or focus groups be recorded or videoed? If so, is permission being sought?

• If any qualitative interviews are recorded, will the tapes be destroyed after transcription?

• Will respondents in qualitative interviews be offered a chance to review the transcripts?

• For quantitative approaches (e.g., questionnaires), will the responses be used in aggregate form?

• How will data from questionnaires be handled? Who will have access?

• How long will the physical copies of the questionnaires be kept before being destroyed (or will they even be destroyed)?

• For both qualitative and quantitative approaches, are there any questions being asked that might be seen as offensive to some participants?

• Where will the data collection take place?

• Do you have permission from the relevant individuals and/or organizations to collect data from particular locations?

• Will the respondent be required to give his or her name? If so, how will this be handled during data storage *and* in the final reporting of the data analysis? If the respondent agrees to have her or his name mentioned, signed consent must be given.

• How will issues of anonymity be addressed if a respondent does not agree to have her or his name used? Will it be possible to identify individual respondents in your write-up?

• In general, will respondents be assured of anonymity? If so, how will this be achieved?

- Will a cover sheet be used for questionnaires? If so, is it attached?
- How will informed consent be given? In other words, how will each participant be made aware of the research project, their role in it, and how the results will be reported?
- Will the respondent be given the opportunity to opt out of the research at any stage?

This is just a small list and, of course, other issues may be relevant to your particular research. Many of the main issues here are applicable to a range of methods. However, as we have noted above, each method – in particular newer, emergent methods – raise distinctive issues which you have to consider. For instance, something as apparently innocent as taking or using (still or moving) pictures of people may require their informed consent. Also, you have to be mindful of the (geographical) context of application, especially if you are contemplating overseas fieldwork: ethical codes are not universal around the world and they vary, as do the expectations of you as a researcher.

- *The questions (or the nature of the questions, if qualitative)* Most ethics officers or committees will want to see a list of the questions you plan to ask, or at the very least the nature of the topic areas if a semi-structured or unstructured interviewing technique is being used. Often, this is the only way they are able to assess qualitative research, so being accurate in the application is critical. For quantitative research (such as questionnaires), they are likely to ask to see the final questionnaire (which may be another consideration when you are preparing your research proposal).

The consent sheet and the information letter: qualitative research

It is not uncommon for ethics officers or committees to require the use of an informed consent sheet that the participant/respondent actually signs to acknowledge consent. As part of the scrutiny process, they may wish to view, and require you to revise this based on feedback before its use is authorized in your research. This document is very valuable: once signed, it becomes evidence of the participant's informed consent and it should be kept safely and securely. A sample consent sheet is presented in Box 9.2. Note that it contains the four critical issues raised earlier with respect to the respondent's 'rights' during the course of the research.

Similarly, you may be asked to provide an information letter to each respondent (in addition to the consent form). This should contain all the information about your study so your potential respondent can make an informed decision about whether to consent to participate. An indicative example is below (Box 9.3).

Note that the detail provided in the actual ethics application usually does not get copied over to the information sheet. The reason is that respondents do not need to know the intricate details of, for example, the methods being used in the research. They need to know briefly what the study is about, how they were selected, why they were selected,

Box 9.2 An example of a signed informed consent form

[SHORT TITLE OF YOUR PROJECT]

CONSENT FORM

I have read the information sheet concerning this project and understand what it is about. All my questions have been answered to my satisfaction. I understand that I am free to request further information at any stage.

I acknowledge that:

1. My participation in the project is entirely voluntary.

2. I am free to withdraw from the project at any time without any disadvantage.

3. The data will be destroyed at the conclusion of the project but any raw data on which the results of the project depend will be retained in secure storage for five years, after which they will be destroyed.

4. I understand that an 'open question' technique may be followed, where the questions asked of me may not have been predetermined and will depend on how the interview develops.

5. I understand that there is no risk to my participation.

6. I understand that I shall not be remunerated for my participation in this project.

7. The results of the project may be published and available in the library but every attempt will be made to preserve my anonymity.

I agree to take part in this project.

(Signature of participant)

(Date)

Source: authors

Box 9.3 A sample information letter

[SHORT TITLE OF YOUR PROJECT]

INFORMATION SHEET FOR PARTICIPANTS

Thank you for showing an interest in this project. Please read this information sheet carefully before deciding whether or not to participate. If you decide to participate we thank you. If you decide not to take part there will be no disadvantage to you of any kind and we thank you for considering our request.

What is the aim of the project?
[IN NON-ACADEMIC LANGUAGE, EXPLAIN BRIEFLY WHAT YOUR PROJECT IS ABOUT]

What type of participants are being sought?
[EXPLAIN THE PROFILE/S OF PEOPLE OR GROUPS OF PEOPLE WHO ARE THE SUBJECT OF YOUR PROJECT]

What will participants be asked to do?
Should you agree to take part in this project, you will be asked to . . .

[WRITE A CLEAR EXPLANATION IN NON-ACADEMIC LANGUAGE OF EXACTLY WHAT PARTICIPANTS WILL BE ASKED TO DO, INCLUDING HOW MUCH TIME IT COULD TAKE. IT IS HERE WHERE YOU WOULD MENTION ANY RISK TO PARTICIPANTS, IF ANY]

Please be aware that you may decide not to take part in the project without any disadvantage to yourself of any kind.

(continued)

Can participants change their mind and withdraw from the project?

You may withdraw from participation in the project at any time and without any disadvantage to yourself of any kind.

What data or information will be collected and what use will be made of it?

[DISCUSS WHAT DATA WILL BE COLLECTED, i.e. INTERVIEW OR FOCUS GROUP TRANSCRIPT]

The results of this project may be published or made publicly available but any data included will not be linked to your participation. Every effort will be made to preserve your identity. You are welcome to request a copy of the results of the project should you wish.

The data collected will be securely stored in such a way that only those mentioned below will be able to gain access. The raw data will be retained for five years, after which it will be destroyed.

What if participants have any questions?

If you have any questions about our project, either now or in the future, please feel free to contact either:

_____

[NAME OF STUDENT]   or   [NAME OF ADVISOR]

Dept _____   Dept _____

Tel. _____   Tel. _____

Source: authors

what they are being asked to provide, and how the information they provide will be used by the researcher. It should also indicate their 'rights' as a participant/respondent. The entire document should be written in plain English, not in academic jargon.

The covering letter: quantitative research

The nature of questionnaires is that, quite often, the researcher and the respondent never come face-to-face. In these situations, informed consent can be acquired by providing a covering letter to the questionnaire that, not unlike the information letter in qualitative research, contains a simple outline of the project, how the information provided will be used, and the 'rights' of the respondent. A sample covering letter is provided in Box 9.4.

Box 9.4 An example of a covering letter based on previous student research

Dear Sir/Madam,

My name is John P. Student and I am currently completing my undergraduate dissertation at the Department of Tourism, University of Southland. The subject of my research is day-trip travel from Invercargill. A recent New Zealand travel study defined day-trip travel as *'travel of at least 40 km one way from home (or travel by aeroplane or ferry service), outside the area in which respondents usually live or work in day by day for a period of no longer than 24 hours'.*

That study is one of only a few that recognizes the importance of day-trip travel. This lack of research into New Zealand's domestic travel market and, in particular, day trips, is why I am investigating day excursions. I have chosen Invercargill as the base for this research due to its growing population of both residents and visitors, and because it is regarded by many as a useful base from which to travel around the South Island.

I would like to invite you to participate in my research by completing the following questionnaire. Your participation is a vital part of my study and it is much appreciated. **Any information that you should provide is strictly confidential, even though the results of the study will be publicly available in the form of a dissertation/article/book/book chapter. Results will only be presented in aggregate form and will only be used for the purposes of this specific research project.** This project has been approved by the Department of Tourism Ethics Committee. It is being supervised by Dr A.C.A. Demic (Tel. _____).

The person in your household aged over 15 years of age with the next birthday should complete this survey. Once the survey is completed, please mail it back in the enclosed postage-paid envelope.

Respondents who complete the survey fully and return it by **2 November** will be eligible for entry into a prize draw. If you wish to be entered please enter your name and address details into the space provided at the end of the survey. Thank you for your participation!

Yours sincerely,

John P. Student

Source: authors

The chapter at a glance

The main learning points of this chapter are that:

- Ethics are important in the conduct of your research and under no circumstances should you ignore them.
- Ethical issues are present throughout the lifecycle of your research, and your consideration of them should not be limited to data collection and reporting.
- Your research will present you with some tricky dilemmas which you should think through carefully and regularly revisit as you conduct your work.

Dissertation checklist

Before you go further in your work, check you:

1.	Read and understood both your institution's policy and programme guidance on research ethics.	
2.	Asked for guidance if you did not understand your ethical responsibilities to your research subjects, advisor/s, institution and other stakeholders.	
3.	Documented the full array of potential ethical issues that may arise as a result of your research.	
4.	Devised measures to ensure the ethical issues are adequately addressed.	
5.	Entered an ethical assessment within your research proposal.	

Part III

PRODUCTION

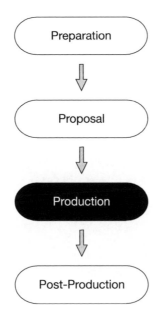

10

OBTAINING AND ACTING ON ADVICE

Learning outcomes

By the end of this chapter you will be able to:

▢ Recognize the different models and levels of support available to dissertation students.

▢ Understand the reciprocal nature of rights and responsibilities in your relationship with your advisors.

▢ Make the most of the support available to you.

Independent research does not have to be isolating

Earlier, on several occasions, the need to seek assistance has been noted. It is true: your dissertation is about your work and your ability to demonstrate your skills and competence as an independent research worker. That said, no person is an island and your work does not – and will not – take place in isolation. It is often helpful, not to say reassuring, to talk through issues with your fellow students because it is highly likely that the sorts of issues affecting you are those that your friends and peers have been or are currently encountering.

A dissertation can be a fairly isolating experience when students feel that they are alone in experiencing difficulties. This is not intended to be the case and, beyond informal support through your social networks, formal systems are in place in every institution. Almost always at undergraduate level one member of staff will be nominated to oversee your research, but there are some instances where committees of academics may be appointed for this purpose (although this is more usual at master's and doctoral level). The aims of this chapter are to develop your understanding of the roles of the staff assigned to support you and, more importantly, how to make the most of what is available to you.

Levels of help: advisors and supervisors

This chapter could have featured much earlier in this book. As we noted in Chapter 3, students often discuss their initial ideas with academic staff. Prior to submitting their research proposal (Chapter 6) they might informally discuss their opinions about the literature (Chapter 4), potential methods (Chapter 5), ethical issues (Chapter 9) and safety issues (Chapter 8).

In most institutions you will be allocated an advisor and you will have little or no choice in this process. Sometimes advisors are appointed before students start their research journey. By and large, though, advisors are allocated based on their fit with the topics proposed by students and/or their availability. In this case they are usually appointed after the proposal has been submitted and, as a result, they may be asked to assess it (Chapter 6).

In this book we have assumed that you will face the latter situation, and hence there are quite instrumental reasons for approaching this topic later in the book, namely: to encourage you *not* to depend unduly on advice from academic staff; to learn to trust your judgement and make appropriate decisions for yourself; and to take ownership of your own project as far as possible. This should make for a much more rewarding learning experience.

The terms 'advisor' and 'supervisor' are often used interchangeably and they can be interpreted in a number of ways. Most institutions publish their definitions of the level of guidance that students may expect to receive during independent research. They should be articulated in the handbook or other guidelines provided by your institution for your dissertation. One important first step is to make sure you read these so that, for instance, you understand who is available to help you, when, how much and what type of assistance they are able to offer you and at which point/s in the course of your work.

In Chapter 1, we noted broad differences in interpretation between 'advisors' and 'supervisors' in the level of assistance they are able to provide. For reasons of fostering your skills as an independent researcher, we have used the term 'advisor' in this book Nevertheless, we can summarize these differences in the following working definitions which you can compare with the labels used by your institution:

- *Advisors* usually provide advice and help in a more general way.
- *Supervisors* usually take a closer view of your work and have a more detailed academic role in guiding your dissertation.

Advice can of course be accepted or rejected. However, supervision implies greater compulsion through instruction which it would be inappropriate to discount. In general, supervisors tend to work more closely with students and offer more regular, detailed or direct input which may involve reading parts of the draft of your work. In addition, they may offer specific technical instruction on particular methodological issues. In contrast, advisors may meet students more irregularly, perhaps at certain key moments in the process, in order to help them overcome key obstacles along the route of their research. For example, they may arrange meetings to give feedback on your proposal and to monitor progress at regular milestones along the way. Their role may not include reading drafts of chapters but they are still likely to give advice on technical aspects. Within the relationship the emphasis is on you as a student to reflect on their comments and make your own appropriate decisions on how to proceed. Indeed the first formal contact you have with your advisor/s may be when they meet you to feed back on their assessment of your proposal. Marking pro formas (see Box 15.1) offer your advisor/s little scope to give in-depth comments in writing, so their 'qualitative feedback' and comments at that meeting are especially important to note and reflect upon.

Whatever the specific terms used in your particular institution and the roles attached to them, these academics are a main source of assistance and, if their support is used wisely, they are potentially a key success factor in your research. However, the relationship between students and advisors is a variable one, and it is often punctuated by misunderstandings between both sides. As several commentators have observed, there is 'no right way to supervise a dissertation student' (Woodhouse 2002: 137; Exley and O'Malley 1999) and students report a wide range of experiences in terms of the support they have received. As this relationship is of critical importance to your work you need to do your best to make this working relationship function to your advantage and it is to this we now turn.

Making the most of your advisor

At the outset you should recognize that your advisor is not there to answer every simple, little question you may have. You may think that your advisor is on-call all the time and just waiting to answer numerous e-mails from you. This is not the case and you need to make every contact a meaningful one.

The central point is that *your* dissertation is *your* piece of work – you are the one who has been trained, who is conducting and being examined in the research process. The academic/s assigned to oversee your work are not being examined for the volume or quality of their insights. In short, their (main) responsibility is to ensure that you are able to complete the research and submit material for examination.

It is also worth remembering that advisors are not the font of all knowledge, however eminent they may be or how much they may like to think they are. Indeed, as advisors they may be familiar with the general subject area rather than being experts in the specific theme you are researching. For instance, adopting the example from Chapter 3, they may not be an expert in the niche of tourism and disability, but they may have widespread expertise as eminent academics in the area of tourist behaviour.

As such, your advisor/s may only be able to provide hints and tips on the specialized topic you have chosen to research (Chapter 3). In this instance, the real experts are those academics whose journal papers or research reports you will have consulted in constructing your literature review (Chapter 4). If you have a detailed question about a particular aspect of a paper you have read that is of special significance to your work, there is no reason why you should not contact these people directly. Most academic papers give the e-mail contact address of the ('corresponding') author/s and some scholars are happy to answer *specific* questions about their work. Let's be clear, enquiries as general and vague as

'I am doing a project on . . . , I have seen your paper on . . . , can you tell me about the work you've been doing?' are not going to motivate them to reply.

Turning back to the question of how to make effective use of your advisor/s, we would suggest that before attempting to arrange a meeting you should:

- *Be very clear what you want to see your advisor/s about.*
  Is it a question of procedure? If so, is it covered in your dissertation handbook? If not then go ahead and ask for help.

- *If you are asking advice on a particular academic subject (e.g. a methodological question) make sure you have all the information available.*
  There is no point in arranging a meeting and then arriving with no clear questions or information. This will create a bad impression. You are more likely to get effective help (and even sympathy) if you are prepared and you are able to demonstrate that you have reached an impasse after thinking through and acting upon the issues as far as possible for yourself. The same applies if you are discussing issues by e-mail: make sure you put your concerns or questions clearly and in an informed way. Long, unfocused e-mails are not as effective as clear, sharply focused enquiries.

- *It is important you convey to your advisor/s that you are a well-organized student who takes a good degree of responsibility for your work.*
  There is no value in asking what to do without at least trying to present what you see as the potential options or solutions and their relative merits in the context of your work.

- *Remember to keep your advisor/s informed of your progress and any major changes in the direction of your research.*
  In some institutions regular (i.e. scheduled) progress reports are a required part of the dissertation process. It may well be that in this case you talk to your advisor/s before submitting your progress form. Alternatively, your advisor/s are likely to discuss the contents of the report with you and any issues it may raise.

- *Make sure you are aware of the availability of your advisor/s.*
  Dissertation research can often take place outside regular term or semester times during 'vacation' periods when your advisor/s may be away, perhaps at conferences or on annual leave (and unable to respond to your queries).

What to expect from your advisor/s

Research has shown that in many cases students have numerous expectations of the academic staff supporting their project. For example, Phillips and Pugh (2000) found that students expected them to supervise: what they meant by this was to always be available when they were needed, to read work well before meetings, and to be the source of all knowledge on all aspects of their dissertation.

Table 10.1 shows the range of student expectations. In part, these are not entirely unrealistic assumptions to make regarding the advisor–student relationship, and certainly

Table 10.1 Students expect their advisors to . . .

1	Offer an acceptable degree of supervision
2	Read and comment constructively on their work
3	Be available when required
4	Be friendly and supportive
5	Have a good working knowledge of the subject
6	Have interest in the topic
7	Be in a position to suggest additional information
8	Contribute to the success of the research

Sources: modified from Phillips and Pugh (2000); Woodhouse (2002)

the idea that they should be 'open and supportive', 'able to structure your meetings', and 'to show an interest in your work' are all important characteristics that you should expect. However, in practice it is often the case that expectations escalate, in which case they start to become unrealistic or at best idealistic. In a small-scale study Woodhouse (2002: 140) revealed how some student expectations were misplaced, with one student expecting 'to receive guidance on what to read'. Of course, as we have noted in Chapter 4, this is not part of the advisor's role in that working through the literature is a student's task, and students with three (or more) years of experience and studying a topic of their choosing should be able to generate their own reading. Nevertheless, you can expect some general guidance (i.e. hints, pointers, tips) from your advisor/s if you have started your reading and have encountered some difficulties. This is general help and not of the type where your specific reading strategy will be dictated to you (as it might have been the case in taught modules) nor will they supply you with a 'reading list'. Ideas on how to read are discussed in Chapter 4, and these are the starting points and strategies you should follow.

Another common expectation is that you will receive constructive criticism. If you are merely told something is poor or incorrect without being given some indications of how to correct the problems or shortcomings, then this is of little help to you. To be effective, the feedback you receive has to inform you not only where you are going wrong but why. This is not to suggest your advisor/s should rewrite problem areas, but you should certainly be given advice on how you can improve your work.

What your advisor/s expects from you

Just as you have expectations of your advisor/s, so they have views regarding what you should do. In Chapters 8 and 9 we referred to this as 'research citizenship', and in this regard your institution bestows particular rights to support and guidance upon you. However, you should not forget that you have certain obligations to your advisor/s. The latter tend to include (based on Phillips and Pugh 2000; Woodhouse 2002) the understanding that you will:

- Exercise independence in terms of your work.
- Follow the feedback and constructive criticism that you are given.
- Attend meetings when they are scheduled.
- Produce material for meetings when requested.
- Give an honest account of your progress.
- Show a strong commitment to your work.
- Inform your advisor/s of any major changes in your circumstances that may impact on your progress.

As we noted above, gaps can emerge between the expectations of students and their advisor/s. Table 10.2 highlights these differences. As you can see, advisor/s usually have expectations that you will follow their advice and meet regular deadlines. By comparison, students view the process as being much more supportive and understanding.

Effective advisors will be good communicators and they should make it clear to you at the outset what they expect from you. They will also inform you of the likely nature and frequency of your contact with them, whether by means of (formal) face-to-face meetings and/or by e-mail and through other virtual learning environments. They may mention periods when they are unavailable to you (see above). They will also explain to you the standard of work to which you should be aspiring.

These basics should be established at your first meeting. If this first formal meeting takes place *before* your research proposal is submitted (see above), most advisor/s will expect you to:

- Have a research title and evolved a research problem which you have turned into aims and objectives (see Chapter 3).
- Have ideas for your possible methods (Chapter 5) and likely timetable moving forward (Chapter 7), as vital aspects of getting your research work underway.
- Have given some thought to the possible implications of your proposed work in terms of your personal safety (Chapter 8), research ethics (Chapter 9) and the resources you may require to conduct the research.

Table 10.2 Variations in expectation within the student–advisor relationship

What students expect from their advisors	What advisors expect from their students
• Read work well in advance of meeting and provide written comments	• To be independent even though some aspects demand conformity
• Be available when needed	• Follow advice given, especially when at the request of a student
• Hold structured meetings leading to relatively easy exchange of ideas	• Produce written work that is not just a first draft
• Be constructively critical	• Have regular meetings
• Be friendly, open and supportive	• Be honest when reporting progress
• Have good knowledge of research area	• Be excited about their work
• Show sufficient interest in research to put more information in student's path	
• Sufficiently involved in their success to help get a good job at the end	

Source: modified from Phillips and Pugh (2000); Woodhouse (2002)

- Enquire about assessments or approval mechanisms (such as for getting your research proposal, health and safety or ethics documentation sanctioned).
- Make them aware of any (final) training or distinctive technological needs that have otherwise not been met so far in your programme.
- Ask about anything you're not clear about!

To be clear, you should take the lead in raising these issues and you should not expect your advisor/s to predict the gaps in your knowledge of the dissertation process.

What happens if things go wrong?

Fortunately, the majority of advisor–student relationships tend to be productive ones. Sadly however, there are occasions where they break down. If you are having extreme problems in working with your advisor/s then you need to seek advice on how things can be improved. In extreme cases this will involve a change of advisor/s who may be more in-tune with your topic. These types of problems are usually dealt with by the person who coordinates the dissertation module for your particular degree programme.

In most cases the problems encountered by students can be overcome through sensible strategies and constructive dialogue with their advisor/s. Even the most thoroughly planned research has difficulties at some point, and for dissertations some of the most common that you should be aware of are:

- *Problems of timing* – many students do not construct realistic timetables (see Chapter 3), but this is something on which your advisor/s should give feedback at an early stage. Remember to specifically ask for comments on the validity of your timetable.
- *You have personal problems, such as illness* – again, you should inform your advisor/s if these difficulties are affecting your work. Life frequently gets in the way of work and your advisor/s expect to be kept informed in broad terms. Most institutions have a set of procedures for mitigation and extension, and your advisor/s should be able to inform you of what to do. Take action on this as soon as possible; there is no point in telling people long after the event. Help them to help you – they can only act as a key advocate for you (in examination boards or mitigation committees) if they are aware of what is happening in the background.
- *Writer's block* – you may find it hard, even impossible, to get your ideas down on paper or at the keyboard. You should discuss this barrier with your advisor/s who should be able to tailor their suggestions depending on the difficulties you are encountering. They will also be able to point you in the direction of dedicated study skills teams at your institution. In our experience, some students struggle because they attempt to change from their natural writing style. All of a sudden, they try to write in an ultra-formal, indirect style which is over-reliant on the passive tense. When asked why, they explain this is what they perceive they have read in tourism journals (when actually it is more common in hard science) and they feel that they have to emulate it in their dissertation.

- *The survey you have undertaken proves less than successful and you have a low response rate* – once more, you should discuss survey design and strategy early on in your preparation work, and it is always useful to have a back-up plan, although this can clearly impact on your time management (Chapter 7). This is something you can raise with your advisor/s in your meetings. In other words, you should try to anticipate any potential problems or delays to your work early on by discussing them with your advisor/s and trying to evolve some contingencies. This may, for instance, also include how you approach your personal safety (Chapter 8) and research ethics (Chapter 9).
- *Your computer has crashed and you have lost all your data and/or drafts of your work* – this is a common problem but it should never arise, because you should always keep a back-up copy of all your work (Chapter 14). While this is not really something that should concern your advisor/s because it is your responsibility, they may be able to point you in the direction of assistance as well as to help you formulate strategies for reorganizing your work flow and dealing with the unexpected delays to your project (Chapter 7).

Advisor/s as examiner/s

At many institutions dissertations are assessed by two people, among whom may be your advisor/s (see Chapter 15). Clearly, in these circumstances the relationship between student and advisor is very different. From your point of view it places new demands on how you should understand their role. In particular, you need to be fully aware of the assessment criteria and conventions your institution employs. For instance, if your advisor/s are your examiner/s your institution may have strict guidelines about what, if at all, these staff members are allowed to read of your dissertation work. This may include what type of content (e.g. only literature review and methods but not data analysis or conclusion), what proportion of the text (e.g. one or two draft chapters) and/or at what time/stage in the research process (e.g. during 'writing up', not within two weeks of deadline etc.). Such regulations should be formally detailed in the dissertation handbook or other guidelines supporting your research.

The chapter at a glance

The main learning points of this chapter are that:

- **Advisor/s with clearly defined responsibilities will be nominated to support you.**
- **You should use the resources of your advisor/s very wisely.**
- **Just as your advisor/s have clear responsibilities to you as a good 'research citizen', you have a series of responsibilities to them.**
- **Your advisor/s may also be responsible for assessing your research proposal and your final submitted dissertation.**

Dissertation checklist

Before you go further in your work, check you:

1.	Are aware of the type and level of advice, support and guidance you may expect during your dissertation work.	
2.	Know at what point in the dissertation process (i.e. pre- or post-proposal) your advisor/s will be appointed.	
3.	Have met and discussed your work with your advisor/s at the earliest appropriate point.	
4.	Understand how and when to make best use of the human resources at your disposal.	
5.	Know what to do if things don't work out with your advisor/s.	
6.	Are aware of whether your advisor/s will also act as examiner/s for your research proposal and/or submitted dissertation.	

11

DIRECTING YOUR DATA COLLECTION

Learning outcomes

By the end of this chapter you will be able to:

- Match your methods more precisely with your objectives.
- Deal with several common issues that students encounter conducting empirical work.
- Make more informed decisions on whether your data collection is fit for purpose.
- Compile notes to contribute to a significant part of your methods chapter.

This is not a methods book

There are many practical considerations to bear in mind as you navigate your way through your dissertation journey. As noted previously, the relationship you build with your advisor/s will have a major bearing on the success of your work because you will know who, when and how to ask for advice on key matters as they arise during your research.

Structures like these are certainly very important, but they are no substitute for your ability to make the right decisions at crucial moments based on your prior learning. You will already have made many decisions when you were thinking through the direction your work will take (Chapters 3–5) and when you were preparing your research proposal (Chapters 6–9). Other key moments are ahead of you. These relate to the collection and analysis of your data, which we discuss in this and the next chapter respectively.

To be clear at the outset, and to repeat what we have said earlier, this is *not* a methods book, and this chapter is not intended to be a comprehensive discussion of research methods *per se*. There is a large body of knowledge dealing with methods in a more generic sense in the social sciences, and several detailed guides to methods as they relate to tourism studies (Finn et al 2000; Phillimore and Goodson 2004; Ritchie et al 2005; Brotherton 2008; Buglear 2010; Baggio and Klobas 2011; Hall 2011a; Veal 2011).

It is beyond the scope and remit of this book to discuss the details and intricacies of methods and techniques. Instead, this chapter aims to alert you to a series of common issues that students encounter in the collection of data for their dissertations. In the previous stages of your study you should have been introduced to a range of quantitative and qualitative survey instruments and the associated analytical procedures. Like other parts of your programme, the dissertation should only assess you on skills and knowledge in which you have received training at some point.

Thus, we start from the assumption that you have received dedicated training beforehand, and that what you may lack is the experience of applying your learning in a larger piece of work and/or on your own. As helpful as methods textbooks or lecture courses can be, the particularities of your research can make putting the principles into practice a more difficult task than you may anticipate.

As we have learned in Chapter 7, data collection (and data analysis) requires a significant investment of your time. As such, it is not unexpected that you should want to maximize the return on your investment; that is, to ensure that the ways in which you capture data are the most fit for the purpose (i.e. relevant, efficient and effective) of addressing your objectives (and answering your research questions). In order to empower you to make such assessments, this chapter presents a series of key questions you should pose yourself while you are developing your methodological approach.

Answers to these questions will help you establish and assess continuity throughout your dissertation. They are also vital resources for compiling your methods chapter which acts as an important 'bridge' within your dissertation between why you have conducted the work (i.e. introduction, literature review) and what you have discovered (i.e. results, conclusion). This is why the methods chapter is one of the most pivotal chapter in your dissertation (Chapters 2, 5 and 13): examiners want to be reassured that your research was properly planned, executed and reported. Validity, reliability and replication are critical criteria for evaluating the veracity of social science research. Thus, it is absolutely vital that you should record – and later *explain* – the reasons for the decisions you take, not merely describe them (Chapter 13).

We start by expanding further on connections at a macro-level; that is, between your objectives and choice of methods of data collection. In one sense, some of this material could have been presented earlier in the book in Chapter 5 (on selecting methods) or Chapter 6 (on compiling your research proposal). Some dissertation programmes require quite an extensive treatment of methods as part of the approval process; in which case it would be appropriate to read this chapter before you write your proposal. In contrast, other programmes require more of a methodological sketch within the proposal in order to get a sense of what you may do and hence whether the methods are appropriate to the objectives. They expect you to conduct most of your intensive work on methods development *after* the proposal has been submitted and assessed, perhaps under the guidance of your advisor/s. The clue will be in the word count and approval criteria for your proposal. In this book, however, we have adopted the latter scenario.

Making and maintaining connections through data collection

Given the vast array of potential research methods and techniques in your armoury (Chapter 5), before deciding which method/s to employ in your data collection you must first establish their *relative* merits, either individually or in combination. This means that you need to establish their respective advantages *and* disadvantages, or positives *and* negatives, versus the alternatives available to you.

The final decision as to the method/s you use should be based on the balance of the evidence (not just a blinkered view of apparent virtues). The ability to identify both merits and demerits and to make an informed decision based on an assessment of both sides of the equation is the hallmark of a more critical student.

In starting to make this assessment, five important and immediate questions would be:

1. Are the method/s applicable in the specific context of your research and its purpose?
2. What are the advantages of using the particular method/s, either individually or in combination?
3. What are the disadvantages of using the particular method/s, either individually or in combination?
4. Will the method/s yield the volume of empirical data you require in the time you have available?
5. Will the method/s yield the quality of empirical data you require?

By answering these questions you are taking the first steps towards assessing whether your potential method/s will allow you to address your objectives. Put another way, are they fit for the purpose of your research and have they the potential to result in a successful outcome?

The first question may appear unnecessary but it is not. Too often students are wedded to methods which they have preordained as relevant to their objectives. For instance, we have been confronted by students who aspire to understand *in depth* the meaning and experiences of tourists by the exclusive means of a simple questionnaire survey. Clearly, interviews are absolutely essential to such a goal but they have been (erroneously) dismissed because students are 'more comfortable with numbers' (than with people, it seems).

Answers to the second and third questions are important because they progress your thinking from principles to practical application. You may be aware of the general advantages and disadvantages of particular methods as they appear in the literature, but you may not have thought these through in the particular instance of your research. What might seem like an ideal method 'on paper' or 'in theory' may not be feasible within the parameters of your dissertation. For instance, focus groups are popular among many students at first inspection, but this view is often reappraised in light of the time and financial expense of conducting them properly (Chapter 7).

Another way to establish the most appropriate selection of method/s is to map your method/s against objectives (and research questions if you use them) in a diagram akin to Figure 5.1 presented in the methods chapter, or in an appendix to your dissertation (Chapters 13 and 14). The main advantage of this approach is that it allows you to demonstrate clearly the connections and maintain them within your work. It is especially helpful if you are using multiple or mixed methods, and it can act as a useful prompt when you come to write up your results (Chapter 13): it reminds you where to look for results as they relate to your particular objectives (or research questions).

In a basic sense, such an audit also helps you to appraise the *extent* to which your objectives (and research questions) are adequately covered by your choice of method/s. Research methods may be legitimately applied to generate data in order to address more than one objective (or research question). If an objective is covered by multiple methods, ask yourself:

- Is there any risk that you may be concentrating too heavily on that objective?
- Is this at the detriment of other objectives (i.e. you fail to collect sufficient data on them)?
- What are the analytical challenges in terms of interpretation and reporting?

Alternatively, if an objective is less well covered, you should be asking yourself:

- What other method/s might be successfully employed to address it?
- Do you, in fact, need the objective?
- Should you de-emphasize it?
- Does it require rewording?

With respect to the latter point, it is important to note that aim/s and objectives are not set in stone. There are occasions – especially towards the end of the process (Chapters 13 and 14) – when it may be necessary to revise objectives during writing up. However, your objectives may be adjusted in the course of devising, testing and conducting your empirical research. Put another way, the objectives you state in your research proposal do not have to be precisely the same as those you enter in the final document. Clearly, it is desirable that they are as close as possible to one another and that you do not revise them too frequently. Not only is there an opportunity cost in terms of lost time but also it may be more difficult for you to keep track of your current objectives. As such, you may run a greater risk that the survey instruments do not necessarily generate appropriate data and results when compared to your final objectives (Chapter 14).

Returning to this auditing approach, the main limitation is that, unless you use some sort of coding system, such a diagram just establishes where the links are between methods and objectives, rather than the relative importance of the method/s in contributing towards the delivery of individual objectives.

To overcome this, a further simple device – a table of objectives versus methods (Table 11.1) – can be used for this purpose. By means of a symbol – here a 'tick' – it is clear which objectives are being addressed by the three methods that the student is using to investigate the objectives for the hypothetical dissertation introduced in Box 3.3.

The larger ticks denote the methods that are more appropriate to, or have greater relative importance in the delivery of, particular objectives (i.e. they are anticipated to generate the highest quality and/or volume of data); conversely, smaller ticks imply methods that will contribute less information.

Table 11.1 Tabulating objectives against possible methods

Objective (from Box 3.3)	Method		
	Questionnaire	Interviews	Observation
1. To examine the levels of participation and exclusion of people with disabilities in holiday-taking	✔	✓	—
2. To investigate the nature of holiday decision making by people with disabilities in terms of perceived and actual barriers	✓	✔	✔
3. To understand the meaning of holidays to people with disabilities and their families	—	✔	✓

Key: ✔   major method for this objective
      ✓   secondary method for this objective
      —   not using this method to address this objective

Source: authors

A questionnaire survey is identified as the main means by which to establish the levels of participation in holiday-making, primarily by means of establishing a benchmark (i.e. proportion) within the sample. Interviews are used in a secondary regard to explore whether there are any underlying factors behind the observed participation/exclusion rates. With regard to decision making, the student thinks that the questionnaire may, in basic terms, allow the identification of those who commonly contribute to, and drive forward, decisions about holidays. However, the richest seam of data is to be obtained via semi-structured interviews. Observation at attractions offers a further significant opportunity to gather data and develop ideas on-site: when holidays are being taken rather than at home before or after holidays. Finally, the student considers that the meaning of holidays can only be articulated via interviews, although the observation of holiday experiences as they happen may contribute some insights.

Auditing questions

Mapping your objectives against your methods may seem like a tedious exercise but it serves two important functions: first, it forces you to consider more carefully the relevance of your proposed empirical research; second, and connected to the first point, it makes you think through the design of your work more precisely.

In practical terms, it is a first step towards minimizing the risk of your wasting time and money. Clearly, the last thing you want to do is employ methods that are inappropriate to your topic. This is also true at a micro-level. The questions you ask, for instance, within a questionnaire survey or as part of a semi-structured interview, have to serve a purpose

Table 11.2 Seven questions for auditing your survey instruments

Question	Ask yourself this
1	What is the purpose of each question – why am I asking it?
2	What objective/s (or research question/s) will each question serve?
3	Have I got too many/few/the right number of questions in order to be able to address my objective/s (or my research question/s)?
4	What information do I hope to generate from each question?
5	What are the likely problems in asking particular questions (in the current way)?
6	Is there a better way of posing the questions/capturing the necessary information?
7	What analytical procedure/s will be applied to the data that each question will generate?

Source: authors

in addressing your objectives. Otherwise, why ask them? You will be wasting your and your respondents' time in generating at best tangential, at worse entirely spurious, data.

In order to reduce this risk, it is useful to scrutinize your survey instruments intensively. A set of seven questions is helpful in this regard (Table 11.2). This scheme is probably best used when you have completed a first draft of the survey instruments you are proposing to use. However, you can also use them in an iterative sense during the drafting process.

The basic premise – which is to get you to be more critical of your approach in a very detailed way – means that you can adapt these questions for all manner of methods. For instance, in the case of focus group research you may not be just interrogating the purpose of questions. Rather, you may also question the appropriateness of prompts or group exercises to guide the discussion. With respect to Questions 3 and 6, you might come to the conclusion that you have too many questions, that the discussion guide may be a little boring or uninspiring, and that there is a risk that your group/s will not yield fruitful, relevant data if you don't take a more creative or interactive approach.

As Questions 4 and 7 reveal, 'downstream' considerations are important, and during the design of your data-collection methods it is worth considering whether your questions are going to generate the type/s of data to which you wish to apply particular analytical techniques. Put another way, are there particular questions, formats or design features that you need to use in order to apply your preferred procedures later? For instance, if you are using a questionnaire, what scales of measurement are being used and hence what kinds of data will each question yield? As you will probably recall, there are four scales of measurement:

1. *Nominal* data are also termed 'categorical data', and they classify cases in terms of concept (e.g. gender, ethnicity, occupation). There is, as a result, little that can be done with these data except categorize them or use them as a filtering device (Bryman and Cramer 2011).
2. *Ordinal* data imply some form of order (e.g. from highest to lowest, best to worse, top to bottom), and not surprisingly they are often used to classify people's

perceptions (Brotherton 2008). Field (2009) uses the example of gold, silver and bronze medals as an example of this ordering, as well as to suggest that we have no indication of the exact level of difference between categories.

3. *Interval* data are measured on a scale where the intervals are equal. Some commentators even use the example of a five-point Likert scale from 'strong agree' to 'strong disagree' as an example explaining that there is a fixed interval of perceptual difference between points on the scale (Brotherton 2008).

4. *Ratio* data have an absolute zero and continuously extend to infinity (Brotherton 2008) and are an extension of interval data in the sense that the ratios between the intervals are meaningful (Field 2009). For example, a room rate of €100/night is twice as expensive as a room of €50/night which is three times cheaper than one at €150/night.

The scale of measurement can have significant implications on the types of statistical tests you may undertake after the data have been coded and entered. Bryman and Cramer (2011: 71) view interval and ratio data as 'the highest level of measurement'. This is 'because there is more that can be said about them' and it is possible to employ 'a wider variety of statistical tests and procedures' with such variables (Bryman and Cramer 2011: 72). Within inferential statistics it is possible to distinguish between:

•   *Parametric tests* – which are based on the normal distribution and require four basic assumptions to be met (see Field 2009) of which the two most important in the context of the current discussion are interval or ratio data and independence (see later in the chapter); and

•   *Non-parametric tests* – these are so-called 'distribution free' tests because they do not assume that the data conform to the normal distribution (Field 2009) but that they are able to handle nominal and ordinal data (as long as they were randomly generated).

Some of the more commonly applied parametric tests and their non-parametric equivalents are presented in Table 11.3.

Questions 4–6 deal with how effectively your questions capture information that addresses your objectives. For instance, could the information being requested be considered to be of a sensitive nature? Could respondents be inclined *not* to answer? What is the likelihood of non-response for particular questions?

Table 11.3 Some common parametric tests and their non-parametric equivalents

Purpose – i.e. to detect differences between	Parametric test	Non-parametric test
Two independent samples	T-test	Mann-Whitney U Test Chi-Square Test for Independence
Sample and population	Student's t-test	Kolmogorov-Smirnov Test Chi-Square Test for Goodness of Fit
Paired samples	T-test for tied samples	Wilcoxon Rank Sum (W) test
Multiple samples	Analysis of Variance (ANOVA)	Kruskal-Wallis (H) Test

Source: authors

Questions 1–3 raise basic issues of efficiency within your data collection. Obviously, there is no point asking more questions than necessary, but efficiency is a complex consideration. In order to address your objectives, you may need to ask several more precise, focused questions within a survey or programme of interviews in order to arrive at the data you seek (Table 11.4).

For instance, one of your objectives may be to investigate whether there is a gender relationship in the commercial success of small- and medium-sized tourism enterprises. To do answer this, you will need to ask at least two specific questions within a questionnaire survey: one dealing with gender of the manager; and another regarding the commercial success of the business, however you chose to define this. One obvious parameter would be annual turnover, but you may think that respondents will be reluctant to provide this either in absolute terms or to indicate it on an interval scale. However, there may be other indicators that you may want to consider such as the number of employees (full-time and/or part-time), (average) occupancy or even the (average) room rate in order to generate a more complete impression of this relationship and/or to act as an insurance by providing indirect indexes of success. Thus, you may in fact be required to ask a minimum of two specific questions within the questionnaire in order to address this one specific objective.

As Table 11.4 indicates, a relatively simple table will help with auditing questions. The first column simply contains the number of the question as it appears in the questionnaire, an interview schedule or topic guide for a focus group. The second column contains the question as it appears and as you propose to ask it (or as shorthand). Within the third column you set out the objective the question is intended to serve, while the fourth column contains an explanation of the reasons for asking the question. These may include the following, either individually or in combination:

Table 11.4 A possible framework for auditing the content of a questionnaire survey

Question	Questionnaire	Objective/s?	Notes (including reasons for/purpose of question)
1	Your gender?	—	General background/explanatory (demographic) variable
2	Your age?	—	General background/explanatory (demographic) variable
3	Size of business?	1	General background/explanatory (demographic) variable. Have used general classification scheme used by EU. Could be considered a surrogate of success
4	Turnover of business?	1	Use class intervals rather than request figure outright. Might not answer question directly
5	What proportion is repeat business?	1	Surrogate variable. Argued to be an indicator of success for accommodation providers by Bloggs (2011). Indirect measure. Demonstrates strength of recurrent demand. Again class intervals to encourage higher response rates

Source: authors

- *Comments on the purpose* – you may want to expand on the particular reasons why the question connects with a particular objective/s or, if the question addresses more than one objective, the relative significance of the question to each.
- *Notes on wording* – academia is full of jargon and technical terms, especially to explain theories and concepts. However, in order to make your research under-standable to your participants you may have to translate some into plain English. As we discuss below, the social sciences are littered with so-called 'fuzzy concepts' that are multiply and variously understood by different stakeholders. As such, questions may have to be drafted to encourage common understanding.
- *Connections to the literature* – as we noted in Chapter 4, your dissertation may be inspired by issues raised or results presented in extant studies. You may wish to apply particular models, concepts, theories or techniques. Perhaps you may be using a standard scale accepted in the extant body of knowledge or you have devised your scales based on the concepts and constructs that appear most frequently in a field, as revealed by a meta-analysis (see Chapter 4).
- *Methodological notes* – you may wish to note why a particular type (e.g. Likert, Semantic Differential) or size of scale (e.g. 3-, 4- or 5-point) was used, or your position on the use of 'neither agree nor disagree' or the inclusion of 'don't know' (Ryan 1995).
- *Potential analytical techniques* – decisions you make regarding the manner and type of data collection can drive the selection of analytical techniques you may employ in the end. The reverse is also true: if you propose to conduct particular types of procedures, you will need to ensure that you collect data that fit their criteria for use. In this respect, issues of sample size or the scale of measurement may be important.

You can tailor the table structure as you see fit. More columns can be added, perhaps one for each of the issues covered by the bullet points above. What is more, although it works especially well for questionnaire, interviews and focus groups, this sort of device can be modified and used with other methods.

In effect, this sort of table serves as an extremely useful record of the reasons for taking particular decisions, and to recall later when you write up your methods chapter (Chapter 13). An effective record mitigates the challenge that there may be several weeks or even some months between the design of research instruments, the collection of your data, and your writing up. Whatever you do, it is important not to rely on memory alone, because you have many decisions to make along the way and why risk forgetting key things? This is not the same as saying that you must use this sort of device in the main text of your dissertation – you may not want to! If you do, you might consider putting it in an appendix (Chapter 14).

Asking more effective questions: questionnaires

So far this chapter has focused on ensuring that your methods are fit for the purpose you intended, and that in the process they contribute to greater continuity within your dissertation. In the preceding section we introduced the idea of efficiency in your survey instruments; that is, ensuring that you only ask the questions you need to in order to be able to address your objectives (or answer your research questions) satisfactorily.

Efficiency, in terms of the number of questions you ask, is important because the time available to collect data is finite. What is more, your respondents have a limited opportunity to be able to take part in your research and, in general, there is an inverse relationship between time required of them and their willingness to participate. You need to strike a balance between parsimony and asking what is necessary for your research.

Efficiency is connected to effectiveness. One other very common means of introducing inefficiency is to pose poorly worded, ineffective questions that, because of their inadequacies, either do not result in the anticipated data or generate spurious information. Inefficiency can be compounded by asking multiple questions on the same theme, perhaps because of vague or tentative wording. As a first step to asking more effective questions, you should consider your answers to Questions 4–6 in Table 11.2.

Question 6 challenges you to examine the mechanics of questions in some detail, in particular the wording and syntax of questions. An immediate issue is whether each specific question will make sense to your intended respondents in the form it is currently written. On one level, this means ensuring that you use the correct language and vocabulary to convey ideas and concepts to the respondent (see below). On another, it is about ensuring the right syntax is used. This is not necessarily about using perfect grammar. Rather, it is about constructing the questions so as to ensure your target audience will be able to understand them and respond as required. For instance, grammatical rules concerning good style and the use of relative pronouns may be suspended because this may make your questions less accessible to your respondents.

From a technical perspective, several shortcomings routinely blight student questionnaires:

- *Missing instructions* – you may not think that it is necessary to include 'Tick all that apply' or 'Tick one box only', and that your respondents will intuitively understand what they need to do. Alternatively, you may be struggling with space and formatting. However, it is important to make it clear to your respondents what they need to do on your behalf.
- *Incorrect filtering instructions* – obviously there is no point asking your respondents to complete parts of the questionnaire that are not applicable to them. This results in spurious data. However, while *most* students remember to include filter instructions to guide, as necessary, their respondents around the questionnaire, you should also remember to adjust the instructions as you revise the questionnaire (e.g. by cutting and pasting to change the sequence of questions).
- The '*and problem*' – one of the most common ways in which you can confuse your respondents is to use multiple operators in a question. As noted previously, by using

the word 'and' in a statement, the respondent is being forced to consider the level of agreement with respect to both clauses in the statement. If it were used in the final questionnaire and data analysis, it would not be clear to which part of the statement the reported score for each respondent referred.

- The '*or problem*' – the use of 'or' introduces a similar dilemma. It is unclear to which part of the statement a respondent's score relates for a statement including two clauses as alternatives.

- '*Fuzzy concepts*' – in a landmark paper, Markusen (2003) argued that recent developments in the social sciences had been accompanied by greater conceptual complexity and the emergence of so-called 'fuzzy concepts'; that is, alluring ideas that seemingly have some explanatory power in the world beyond academia but which defy straightforward definition and hence are more difficult to research. For instance, terms like 'innovation' are popular in academia, policy and practice, but there are often quite different understandings and emphases even within the 'tourism industry' (Coles et al 2009). Some students overlook this and assume there are established, common understandings among users. One approach to 'fuzzy concepts' is to use multiple questions to ensure adequate conceptual coverage; another approach is to define key concepts within the text to ensure that all participants share the same understanding for the purpose of responding.

Of course, the piloting phase of your research should expose many of the shortcomings in your questionnaire and provide you with an opportunity to eradicate them (Box 11.1). So, it is a false economy to skimp on piloting in your timetabling (Chapter 7). If they are not detected at an early stage, apart from potentially erroneous data, poorly designed questions may also result in non-responses. Quite simply, if your respondents don't understand your questions or how they work, they may not answer them.

Non-responses reveal themselves in the coding of your questionnaire responses and, subsequently, the level of non-response can be exposed by running simple frequencies in software packages like SPSS. Questions that yield lower (relatively speaking) levels of response than might have been anticipated based on the total number of (valid) respondents may have been more problematic to understand and/or answer. In extreme circumstances, they may prejudice your ability to undertake particular types of statistical analysis. Not only is it worthwhile doing some follow-up research in this respect, but it is also important to discuss the relative coverage and reliability in your methods chapter (Chapter 13).

Asking more effective questions: interviews

Qualitative research is of course a much more fluid and iterative experience, but it is useful to consider issues of efficiency and effectiveness, especially in how programmes of interviews are designed. As we have noted above, many undergraduate students are under the sad misapprehension that interviews are relatively straightforward. If this were the case, there would be no need whatsoever for the excellent array of textbooks dealing with

Box 11.1 Identifying success in piloting survey instruments

Many advisors extol the virtues of pre-testing and the piloting of survey instruments. This is a sound recommendation which applies equally to qualitative and quantitative methods. While there is a plethora of best practice available to consult, many students struggle to make decisions about which individual questions or parts of the survey are *not* working. This is hardly surprising, because they don't have extensive experience in surveying, they are anxious not to make a mess of their work, and they feel under pressure to 'get it right' if not first time, then pretty quickly. This is exacerbated because they don't have a clear view of what success might look like.

RAG analysis (Red-Amber-Green, after traffic lights) is a highly visual managerial tool. One of its virtues is that it communicates assessments clearly and simply – audiences get the point quickly. In discussions with your advisor/s, you might think of presenting the results of your testing using the RAG 'scale' and a spreadsheet by cross-tabulating the question number (rows) against pilot respondents (columns) and entering data for how well the question was answered in each cell.

For instance, on a questionnaire, if a question was perfectly answered by a particular respondent, enter green. Conversely, if it was not answered at all (and should have been), enter red. If mistakes or misunderstandings were made (e.g. if multiple boxes were ticked when only one should have been or they should not have been completed as a result of a filter instruction), it is clear the respondents proceeded even if they should not have, in which case enter amber. An entirely clean bill of health will be awash with green. This is unlikely first time as is a very angry looking spreadsheet! However, the more 'warm colours', the more work to refine the questionnaire you will need to do (or the harsher a critic you are).

In all likelihood, if you have been careful in your preparation, you will be faced with a predominantly green spreadsheet punctuated with some amber and very little red if none at all. You then decide on action: for example, you may decide that: questions answered correctly 90 percent or more of time require no revision; 75–90 percent of the time require minor adjustments; and less than 75 percent of the time require a major rethink. These thresholds can be adjusted based on the scope or complexity of the questionnaire structure and/or the questions you ask (i.e. easier, shorter questions should perhaps have higher thresholds).

If you have interviewed respondents in the pilot, you should have a fairly shrewd idea of the shortcomings and what changes to make. As a result, you should be able to make recommendations to your advisor/s and focus your discussions on the questions that really puzzle you. Finally, be realistic in drawing revision to a close. People still make mistakes in completing otherwise thoroughly tested questionnaires. This is a limitation of the technique about which you will no doubt want to write about later.

Source: authors

the intricacies of qualitative methods in their various guises. If it were so easy, the former mayor of Benidorm would not have been so effusive in an interview conducted by an experienced newspaper journalist on the historical development of the resort (Hickman 2007). Similarly, there would not be routine discussions about the difficulty of interviewing 'business elites' (Mosedale 2007; Coles et al 2009) and the strategies necessary for eliciting information from such a conspicuous group in society (Harvey 2011).

In some schools of thought on interviewing, systemic preparation of the type described in this chapter may be eschewed: it is somewhat at odds with the purpose of interviews to stimulate dialogue, and it seemingly limits the potential for the 'interview process' to take the discussion – and hence collection of data – into altogether new and perhaps previously unanticipated areas. Simply put, unscripted interviews are the 'purest' form, and as such scripted questions should be no more than departure points – and the fewer the better. For experienced and/or confident researchers, the potentials are exciting and – depending on the nature of the collected data – they may require a post-hoc adjustment of aim/s and objectives.

For less experienced researchers, such as undergraduates conducting research for their dissertations, this is quite a daunting prospect and we would advise you to devise a basic script as an 'insurance' to underwrite your interviews; that is, to employ a semi-structured mode of interviewing. In light of the time you will invest in setting up, conducting, transcribing and analyzing the data, you will be keen to avoid collecting poor quality data that compromises your ability to address your objectives.

A script can help you minimize this risk while offering you the freedom and opportunity to explore attendant issues confidently with your interviewees. Remember, you do not have to ask all the questions on the script, nor in the order in which they are written down. You don't have to be regimented and brittle in your approach, which is the main risk associated with the fully scripted approach. You can allow the interview experience to be iterative and fluid while keeping tabs on whether the topics in which you are interested have been discussed by the interviewees.

In preparing, there are some practical considerations that students overlook which may make both the experience of the interview and the quality of the resulting data more rewarding. First of all, the same issues of clarity, purpose, audience, syntax and vocabulary are just as valid in this context. The use of 'and' and 'or' is generally more problematic in questionnaire surveys than interview schedules. At least during discussions, it is possible for the interviewer to clarify through follow-ups and exemplifications which part of the statement the answer is referring to. A reading of the full transcript of an interview may also clarify the matter. However, it is possible to avoid the problem altogether if your questions are carefully formulated and articulated. Several other issues to consider are:

- *Follow-ups* – you should be prepared to probe issues further, by means of further (scripted) questions, requests for further clarification or exemplification, as well as . . .
- *Examples* – asking an interviewee to cite an instance, case or example is often an effective means by which to render more abstract ideas more tangible. It may also

be an opportunity for you as interviewer to calibrate your interpretations and understandings of key (possibly 'fuzzy') concepts with your interviewees'. Data of this nature may provide you with richly illustrative material for your results chapter (Chapter 13), assuming that its inclusion does not compromise any ethical assurances or standards (Chapter 9).

- *Prompts* – although more commonly used in focus groups, you may want to take artefacts to discuss with your interviewees, such as reports, newspaper articles, results of your own (perhaps from a questionnaire survey).
- *Prioritization* – are there certain questions you would like to ask each interviewee? Are some questions and lines of discussion more important to addressing your objectives? It is worthwhile making a note of them (perhaps by marking them in **bold** or *italics*) and ensuring that you weave them into the discussion.
- *Contextual questions* – don't forget that, in order to analyze the responses, you may need some contextual data to understand the positionality of the response. At its simplest, this may be in the form of basic socio-demographic variables. Sometimes this is collected by a short questionnaire schedule, although it is amazing how often this is overlooked.

Preparation is clearly important, but it is worth remembering that interviews are about giving your interviewees the chance to speak! After all, you have invited them to give their views on issues in which you are interested. They are there to inform you. As Clark (2010) notes, there are two broad sets of drivers motivating people to participate in qualitative research. At an individual level these include interest, enjoyment, curiosity and social comparison, while at a collective level representation, empowerment and informing 'change' are key considerations.

And as a final reminder, don't forget to:

- turn the recording device on;
- take spare batteries with you; and
- ensure you have completed the informed consent forms and other formalities correctly.

Many a time students have failed to get these basic operations right. This will leave you and your advisor/s with some awkward dilemmas to confront.

Drawing an appropriate sample

Finally in this chapter, we discuss briefly the idea of sampling. So far we have concentrated on the practicalities of designing and implementing your survey instruments without discussing in any depth who you are going to access and the strategy you are going to use to access them. Sometimes sampling is falsely assumed to be the exclusive concern of quantitative research, but it is just as important an issue for qualitative research despite the smaller sample sizes that are likely to be used. However, both approaches are intended to paint a picture of the population.

The population may be defined as the 'total membership of a defined class of people, objects or events' (O'Leary 2010: 161). Businesses and organizations may constitute 'objects', but such a definition is important because, in statistical terms, you have to get beyond the lay connotation of a population being comprised of people. The population is the complete set of phenomena and a sample is a sub-set of this group. As with other forms of social science, the ideal situation would be to ask everyone (i.e. conduct a census), but this is not possible in a great many cases (unless the population is very small) and it may not be necessary according to statistical theory. As O'Leary's (2010) excellent discussion makes clear, there are many different types of sampling, including:

- *Random* – all members of a population have an equal and fair chance to be selected for inclusion in the sample and for this reason it may allow for generalization (O'Leary 2010: 167).
- *Systematic* – from a defined population, every nth person is selected in a close approximation to random sampling provided that there are no structural features that result (O'Leary 2010).
- *Stratified* – the population is divided into sub-groups and a proportion of the surveys gets completed in each of the sub-groups. Where the proportion of surveys is equal to the sub-groups' presence in the population, it is described as 'proportionate'. Within each stratum, either random or systematic means can be used to identify individual subjects (O'Leary 2010: 167).
- *Cluster* – this is used where distinctive geographical clusters are known to be representative, and sampling within these clusters is used as a proxy for a more widespread survey of the background population using one of the three strategies mentioned already. This requires an understanding of the characteristics of both the population and the cluster (O'Leary 2010: 168).
- *Convenience* – as its name suggests, the sample is collected in a manner convenient to you as a form of non-random sampling. However, O'Leary (2010: 170–1) takes issue with this term and for her 'convenience sampling has no place in credible research' because 'there needs to be more to a sampling strategy than just convenience'.
- *Purposive* – a catch-all term to include the deliberate selection of subjects because of some important characteristic/s they possess (Brotherton 2008: 172). Ostensibly used in qualitative research, this might mean they act as 'key informants' (i.e. 'handpicked sampling') or that they are chosen by referral (i.e. snowball sampling) or that they are volunteer (i.e. 'volunteer sampling', O'Leary 2010: 170).
- *Quota* – in this case quotas of surveys are completed proportionate to the presence of particular sub-groups in the population. For instance, based on a 50:50 gender split in the population, a quota survey of 300 requires 150 completions by males and 150 completions by females. The process is distinctive for the influence of the selector. Brotherton (2008: 173) refers to this as 'quasi-representative', no doubt because it is intended to reflect the population composition. However, there is not an equal chance of each potential subject in a population being chosen.

There are relative merits associated with each of these strategies, the intricacies of which are beyond the scope of this book (see O'Leary 2010, Brotherton 2008). Because who you sample and how you sample are so important to how you produce knowledge in your dissertation, we would expect you to consult the specialist research methods literature in order to ensure that your knowledge is most up to date. Here we want to highlight two practical issues in terms of operationalizing your research.

The first is the ramifications of particular strategies for the types of statistics you may employ in your analysis. The first four strategies comprise 'probability sampling': this means they are probability based and that there is an equal chance of a potential subject being chosen. As noted earlier in this chapter, the normal distribution is a fundamental underpinning for parametric statistics, and to employ this suite of analytical techniques you must not violate the principle that the observations were drawn independently of one another. In practice, this is equated to ensuring that randomized sampling is used to ensure that systematic connections or biases are avoided.

The second issue is that, although these strategies are straightforward to understand in principle, the selection of your sampling strategy requires careful thought before proceeding because it relates to several other connected considerations, as the examples below indicate. For instance:

• *Dan wanted to talk to tourists visiting a particular city.* His sample would ultimately not be representative of all tourists because he could not feasibly give every visiting tourist an equal chance of being selected. Dan elected to adopt a non-probability sampling strategy. He approached tourists at the local visitor information centre and asked if they would fill out a short questionnaire. He made every effort to 'randomize' the procedure of approaching tourists, such that he would attempt to alternate men and women whom he approached and balance a survey with a younger person with one subsequently with an older individual. While he may have done his utmost to generate what he thought would be a representative sample by means of trying to fill some notional quotas, it is most accurate to conclude that his strategy was convenience based.

• *Haley was interested in surveying local residents' attitudes towards second-home owners in her small town.* She obtained a copy of the local telephone book, which contained in total 24,450 names. Haley knew that she wanted a final sample size of 200, which is enough to allow her to conduct some sub-sample analysis (e.g., how men or women answered a particular question). To send a survey to 200 people (her sample unit) in a town with a population of 24,450 (her sample frame), Haley knew that she would have to select every 122nd person in the phone book. To start, Haley selected a random number between 1 and 122, and that became her starting point. From then, Haley selected every 122nd person. This was time consuming, but it resulted in a probability-based sample obtained via a systematic random sampling strategy. If Haley wanted to, she could have stratified the population into several sub-groups (e.g., particular areas of the town known to exhibit differing socio-economic attributes), resulting in a stratified systematic random sampling strategy. However, unless Haley had access to census data or an electoral register that could be sorted accurately by neighbourhood, this task would have been exceptionally time consuming using just a telephone book.

• *Jose wanted to administer a survey to long-distance walkers along an especially remote track in New Zealand.* His sample frame is the specific track, known to be particularly challenging, thus attracting 'trampers' with high levels of fitness. His sampling unit was users (trampers) on that track. Jose knew that his sampling strategy would be based in principles of non-probability, but he was more concerned about his sample size. Specifically, he wondered whether, when using a convenience-based strategy to obtain a non-probability sample, he should even be concerned about how many people he should approach. While it is true that statistical concepts such as margin of error applies to sampling which is probability based only, it is good research practice to be mindful of principles of robustness and validity even when constructing samples which are based in non-probability. After some thought, Jose decided to survey 50 backpackers. This allowed him to be reasonably comfortable with the validity of the data he obtained.

• *Julie wanted to conduct a series of one-on-one, in-depth interviews with students at her university.* She was interested in their previous experiences on family holidays when they were younger. Julie's work was grounded in anthropological and sociological theory, and she wanted to utilize qualitative methods because she was more interested in obtaining what Clifford Geertz (1973) famously called 'thick data'. For her final dissertation, it was clear that Julie was careful to interview students who reflected the wider sample frame (students at her university). She also utilized a non-homogenous sample approach by interviewing an almost equal mix of older, younger, male and female students. Doing so ensured Julie had a broad representation of views, even though issues of statistical representativeness did not apply in this instance.

The nature of your sample must be defensible, and a thorough review of previous studies will supply you with several precedents on which to draw in making your final decision. To assist with this, you should ask yourself the following questions:

1. Can your sample be a probability-based sample? If so, does cost or time prohibit this?
2. Is your target sample size sufficient for you to conduct appropriate analysis? We have often seen students who rush to collect their data without considering this step.
3. Perhaps most importantly, does your sampling strategy fit with your objectives? In other words, will a non-probability sample deliver data that can answer your research questions?

   What is important to convey in your dissertation, regardless of the sampling procedures you have utilized, is how your sample either helps or hinders your ability to draw conclusions from the data.

The chapter at a glance

The main learning points of this chapter are that:

- You should have been trained on research methods as part of your preparation for independent research in your dissertation.
- This is not a book about methods so you need to look elsewhere for detailed guidance on the principles and detail of data collection and survey instruments.
- Data collection should have a purpose: your empirical data has to be relevant to your aim/s, objectives and any research questions or hypotheses you employ.
- By making connections more carefully in your preparation, you should be able to enhance the efficiency and effectiveness of your survey instruments and data collection episodes.
- In the design of your research tools you should be mindful of the type/s of analysis you will conduct subsequently.

Dissertation checklist

Before you go further in your work, check you:

1.	Have revised your prior learning and training on methods for data collection before commencing this part of your dissertation research.	
2.	Know where to obtain detailed advice on the principles of, and best practices in, research design.	
3.	Considered alternative methods or approaches to addressing the research problem that formulates your work.	
4.	Are able to explain the specific purpose of particular instruments and questions within your research.	
5.	Are ready to communicate clear justifications for how far your methods fit the objectives for your dissertation.	

12

ANALYZING YOUR DATA

Learning outcomes

By the end of this chapter you will be able to:

- More effectively and thoroughly explain how and why you have conducted your analysis.
- Recognize the differences as well as connections between data analysis and data presentation.
- Employ several devices to enhance your data analysis and presentation of your data.
- Articulate the vital difference between results and findings.

Have data, need 'answers'!

Once your data have been collected, depending on the approach you took, you will be faced with either a sizeable dataset in a software package such as SPSS, SAS or STATA and/or a great many pages of data from your qualitative research, perhaps to be handled by NVIVO or ATLAS.ti. It is at this point that many students often throw their hands up in despair and wonder where to begin their analytical work.

In this chapter, once more we adopt the principles that whatever you do – in this case, data analysis – should be consistent with your aim/s and objectives, it should be conducted with clear purpose, and it should be carefully planned. As we noted in the previous chapter, the design of your methods of data collection should be mindful of the types of analytical techniques you would like to use to address your objectives. The mapping of both methods and questions was identified as a key means of establishing continuity. Therefore, it will perhaps come as little surprise that here we argue that your data analysis should be focused and that it benefits from being more tightly choreographed. In this chapter, we present the idea of a 'tabulation plan' or 'tab plan', ostensibly used for, but not restricted to, your quantitative analysis.

With a wide range of software packages available to students, there may be a temptation to 'plug and play' and hope that useful results will reveal themselves or that research questions will be answered automatically. Results do not speak for themselves. What's more, you need to consider data analysis in conjunction with data presentation: how you communicate the data you have generated is important to the overall success of your work. Presentation makes your data accessible to readers (i.e. examiners). As such, some basic principles are discussed for you to reflect upon when you are compiling the results chapter of your dissertation (Chapter 13). Once again, given the nature and scope of this book, we assume you have prior learning on the skills and tools necessary for processing and analyzing qualitative and quantitative data.

Links between sampling, scales of measurement and analysis

Before looking at some of the issues involved in how you can make the most of your analysis, it is worth teasing out further the linkages between sampling and scales of measurement, and their potential impacts on your data analysis.

What is important to convey in your dissertation, regardless of the sampling procedures or surveying instruments you have employed, is how your sample either helps or hinders your ability to draw conclusions from the data. Put another way, has your data collection resulted in reliable and valid data (Knapp and Mueller 2010) with which to address your objectives? The former question is concerned with the consistency of measurement from one sampling episode to another, from one subject to another. The latter is 'usually defined as the extent to which the instrument actually measures' what it is designed to do. While there are several statistical tests that may be applied to more complex survey instruments, you could ask yourself some more basic questions about the number of non-responses, don't knows, incorrect completions, and the number of serial responses within your data (Stapleton 2010). These help you to establish your effective or net response rate. From your total number of respondents, how many completed answers do you have that you can use in the analysis, and is this a cause for concern (or celebration)?

Effective researchers are able to demonstrate their ability to acknowledge limitations in their work and to illustrate the fallibility of procedures (not necessarily through any fault of their own). Do not be afraid to reflect critically on your sample and what the practice of production may mean with respect to your ability to draw conclusions. Often students are reluctant to say anything negative about how their research was conducted. For instance, as you will undoubtedly be aware, non-probability samples generally prevent generalization. We have encountered many cases where students have claimed this of their convenience samples of tourists in particular situations (say, at an airport, on a train, or at an attraction) by erroneously trying to craft the results as if to suggest that they are randomly drawn and hence representative of all tourists in the population.

An inspection of your sample set for any signs of bias that may in some way distort the analysis and/or findings is a vital first step. Such 'sample set biases' can be revealed by a simple inspection of the frequencies of background variables. For example, do they reveal any gender, age or socio-demographic (e.g. educational, income, family-related) skews within your sample? Moreover, are they greater than you would have expected based on what you may know about the background population, perhaps from a secondary data source such as a census? Even if you select participants' addresses for a mailshot from a random number generator, you may still find that your final sample comprises 63 percent women and 37 percent men, not the (perhaps 51:49 percent) ratio you may have expected. Therefore, three connected questions you should be asking yourself are:

1.   Is the disparity between the composition of my sample and the background population (statistically) significant?
2.   Is it the result of a flaw in the design of my sampling strategy? And,
3.   To what extent are my results likely to be distorted as a result?

The third question forces you to consider the possibility that what you subsequently observe is (more) a function of the sampling rather than a reflection of the background population (i.e. it is not necessarily representative but you may still reveal important lessons to be revisited by others in the future).

In many cases such assessments are tricky because no data (apparently) exists or is available to you on the background population (at a sufficiently fine-grained resolution). For example, you may be studying the motivations of visitors to a theme park but the operators are not prepared to make available to you their visitor numbers and profile which they've derived from EPOS data on their postcodes linked to geodemographic profiling software. Alternatively, you may be studying the spending patterns of visitors to a new event, and while you may know the total number of attendees, the organizers have no data on the gender composition or socio-economic background against which you can compare.

Indeed, where there are data on the 'population' ask yourself how they were compiled and whether they are an accurate portrayal. Otherwise, you will be comparing your sample profile against nothing more than another sample (which is often the case in tourism research). This does not invalidate your sample but it means that you have no real way of knowing how representative your (case-study) data are and hence how generalizable your findings may be.

Of course, the discussion above primarily pertains to quantitative methods but there are similar considerations in qualitative research. The way in which interviewees are selected may result in samples that have very distinctive profiles which manifest themselves in their particular responses to certain questions.

Beyond sampling, similar issues relate to scales of measurement. You will recall that, in the case of quantitative methods, questionnaire design requires consideration of the type of data that each question yields (Chapter 11), but this also acts to delimit the array of statistical techniques and indexes you can use. For instance, some measures of central tendency (e.g. mean) cannot be derived from nominal data, and it is arguable as to whether they should be applied with ordinal data (as seems common practice in tourism research). Moreover, this restricts the potential type and extent of inferential testing that can be done (to non-parametric procedures). Therefore, our advice in this regard is simple: as you plan and conduct your analysis, it is paramount that you adopt a conscious effort to recognize and accept the constraints that your data place on the choice/s of technique and procedures you employ and hence the sorts of conclusions you may be able to draw.

Data analysis and presentation

With these words of caution in mind, it is now worth unpacking the relationship between data analysis and presentation a little further. After your considerable investment in generating and encoding your data, it is valuable to spend time reflecting on their role – that is, how they may feature – in your dissertation before you embark on an intensive course of analysis. You should not overlook the fact that data analysis presents you with two connected challenges and hence opportunities to demonstrate your competence as an independent researcher:

- On the one hand, you have to decide what your data mean with respect to your objectives that you set yourself; and on the other,
- You have to communicate to your examiner clearly and convincingly your understandings of the data and their relevance to your study.

By concentrating on what analysis to undertake *and* how to present your results most effectively, you will enhance your chances of compiling a more effective results chapter. There is little purpose in indiscriminately generating as many statistics, tables, figures or even themes and quotes as you can in the vague hope or belief that your objectives will somehow be addressed. It is not the reader's (i.e. examiner's) job to sift through output and to make the connections on your behalf. Throughout the process of data analysis and within your results chapter, the test is not the amount of data you are able to generate, but rather your ability to generate high-quality results that allow you to reflect with authority on your objectives.

Reporting quantitative material

We cannot repeat this advice enough. Many students feel that reporting quantitative results is simply a matter of generating tables or charts in a statistical software package and then inserting these into your document. For some students, quantity – not quality and focus – of data is a form of 'safety blanket'. This is absolutely not an appropriate way to report quantitative data.

There is no 'perfect' recipe for generating quantitative data and reporting them in the results chapter, although you may wish to consider issues of:

- *Appropriateness* – the data you generate should be directed at your objectives (and research questions). Put another way, there is little point producing inappropriate, superfluous data that has no apparent connection to the purpose of your study.
- *Parsimony* – you should only present tables and/or graphics that are going to reinforce the points you make in the text. Often students prepare numerous sizeable tables of data which are then at best only partially referred to, and at worse not mentioned at all with the assumption that readers will make the linkages or draw inferences for themselves.

    Although you should tie your tables into the text of your analysis chapter, you should not repeat most or all of what those tables say in the text itself. This wastes space, not to mention words (which will be at a premium). You should recall that tables are intended to be efficient ways of presenting large volumes of data, so why would you want to accompany a large table with a protracted textual discussion? If you need to comment on table contents you should consider restricting yourself to identifying maximum and minimum values, as well as any trends and anomalies.

- *Efficiency* – some students produce tables *and* graphics to illustrate the same aspects of their dataset twice, in the belief that, by presenting in two different formats, in some way a point will be made more emphatically. For instance, we commonly see tables of gender-based tables of attitudinal or motivational variables accompanied by pie charts as visual representations of the results for individual variables. The latter are clearly intended to add visual impact but end up being little more than padding at best, and at worst distracting and suggestive of a low-level of confidence with statistics. You should only use additional forms of representation (e.g. a chart to augment a table) of the same dataset if they contribute, reveal or illustrate distinct and different points that could not otherwise be made by a single representation (e.g. a table on its own).

- *Simplicity* – some students equate quantitative analysis, the use of statistics and a 'positivist' approach with 'science' and, as a result falsely believe that their analysis needs to be intricate and complex because these are the hallmarks of science. Allied to this is a propensity to over-interpret results when Occam's Razor may be usefully invoked: why look for a complex answer when a simpler one is more compelling?

Box 12.1 shows an example of inefficient or 'over-reporting'. Here a student reports on some basic univariate data derived from a modest survey of German students as return visitors. The pie chart looks pretty proficient and it shows with great impact that a clear majority within the sample had travelled frequently to the UK. It would have sufficed on its own (although pie charts take a lot of space to portray a single variable) as would the table (which is the more efficient portrayal). However, both are not required. Moreover, as we discuss below, were the table to be retained on its own it is debatable whether both the absolute and relative numbers as well as the total columns and the note about rounding are all necessary. As readers, we would have to ask ourselves why the student has done this and what did it attempt to achieve?

Tab(ulation) planning

You should reflect on these issues as you work your way through the analysis of the data you have collected, as well as during the process of writing up (see Chapter 13). For quantitative work, a tabulation plan (or 'tab plan' for short) can be programmed after the questionnaire has been finalized. A tab plan is essentially a map that outlines what is to be done with the data (or, more specifically, each question) once the data collection is complete.

Tab plans, when done before the data collection starts, can actually help to identify gaps in the questionnaire (which may represent limitations of the study for reporting in the conclusion – see Chapter 13) and even any issues relating to potential statistical testing.

Box 12.1 An example of inefficient over-reporting

Table: The number of previous visits to Britain by German students

Visited Britain	Number	%
Once	24	14.8
Twice	25	15.4
Three times	13	8.0
Over three times	100	61.7
Total	162	99.9*

* Based on rounding

Figure: A pie chart of the previous visits to Britain by German students

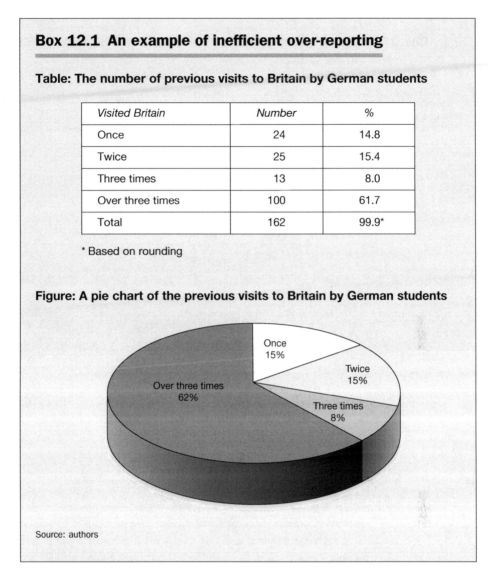

Source: authors

At its heart, a tab plan asks for four pieces of information about each question in the survey instrument:

1. *Who is supposed to answer this question?*
This clarifies who in your sample is meant to answer this question, and hence the effective maximum number of valid respondents (notwithstanding any missing data).

Put another way, it reminds you to look for a skip pattern that respondents should be following. This can be an important question to ask because, when it comes time to do the analysis, it is critical that the responses of only those who were meant to answer the question are analyzed.

In the example depicted in Box 12.2, if a respondent answers Question 1 with 'NO', they are asked to skip ahead to Question 3. If they answer 'YES', they are asked to

---

Box 12.2 Filter questions/skip patterns as a consideration in data analysis

1. Have you ever visited the United States of America?

   ☐ No → *Please go to Q3*   ☐ Yes → *Please continue*

2. If YES, when did you last visit the United States of America?

   ☐ Within the past 12 months   ☐ 1–2 years ago

   ☐ 3–4 years ago   ☐ More than 5 years ago

3. Have you ever visited Canada?

   ☐ No → *Please go to Q5*   ☐ Yes → *Please continue*

4. If YES ...

Source: authors

---

continue. So, the skip pattern only affects those who answer 'NO'. For Question 1, you would want everyone to answer this question because you are interested in learning whether or not they have ever visited the US. For Question 2, however, you would only want those people who said 'YES' in Question 1 to answer. So, when writing out a tab plan, you would allow for this skip pattern to be identified.

2. *How will I analyze this question?*
This is where you start to plan out your analysis proper. The purpose of this question is to think carefully about what kinds of analysis (i.e., what kinds of statistical testing, if any) you will undertake, as well as the metrics you would like to generate and comment upon in the text. For example, in Box 12.2 we might plan, for Question 2, to run a frequency on the proportion who ticked 'within the past 12 months'. We might also want to explore some cross-tabulations that look at how men and women answered this question. We may even want to use chi-square to examine whether there are different gender associations with the US in the data. You may want to note why you are using a particular procedure and not others. Again, this has the added advantage of identifying potential problems and gaps in a questionnaire, especially if it is conducted in the design phase.

3. *What is the base?*
The base refers to how the results of the analysis are presented. For example, let us assume that we are interested in seeing how males answered Question 2 in the hypothetical example above (Box 12.2). When these results are reported, care is taken to alert the reader

that the base is all males in the sample. The purpose of the base is to point the reader, in specific circumstances, to which (sets or sub-sets of) people the representation within our analysis (e.g. a bar chart, pie chart, table etc.) refers. At the tabulation planning stage, establishing the base is critical because it forces you think about what you might be observing about whom.

Of course, we may often decide on more than one base for each question. We may want to see, for example, how Question 1 is answered by males, parents and those with more than one child (either separately or together).

4. *Which objective/s and/or research question/s will this contribute towards?*
Last but by no means least it is worth considering where and how your observations will contribute to the study and its direction. By returning to this question, you will be able to appraise the extent to which individual objectives are covered during your data analysis. Not only does this allow you to examine the breath and depth of the objectives' coverage, but also to benchmark the extent to which you have addressed them versus how far you intended (i.e. designed) them to be addressed. This may prompt you to undertake further analytical procedures to explore, as yet, under-represented objectives or to revisit the importance of particular objectives to the study during your writing up (see Chapter 13).

Selecting analytical techniques and procedures

A tabulation plan can be conducted on its own as a separate task. However, the last point emphasizes that such planning can also be an extension of the table you may have compiled as part of the 'question auditing' which we discussed in the previous chapter.

In light of the wide array of procedures and associated metrics at your disposal, it is important during analysis to keep the units of analysis firmly in mind and to select the appropriate form of statistical treatment. It is in this latter respect that many students – especially those who are less confident with statistics – struggle.

A first point to recall is that there are essentially three types of procedure you can apply:

1. *Univariate statistics* – these are simple descriptive statistics that describe the pattern of observations for single variables.
2. *Bivariate statistics* – these are statistics that describe the nature and strength of simple relationships and associations between two variables (see Table 11.3).
3. *Multivariate statistics* – as more intricate procedures, these data reduction techniques seek to identify the nature and strength of relationships in more complex datasets based on the performance of three or more variables.

Multivariate statistics are often beyond the scope of undergraduate dissertations, but they include 'factor analysis' and 'cluster analysis' which are routinely employed in disciplines such as psychology and management (i.e. marketing) studies. The former is a data reduction technique intended to reveal the underlying structure from a number of variables while the latter is a means of identifying groups of subjects (usually individuals) with common characteristics.

It is, nevertheless, more common to see students use bivariate and univariate statistics in their dissertations. In fact, ironically some students overlook univariate statistics, probably for the reason that they appear to be too simple (and hence are incorrectly perceived as not being scientific enough, or as apparently incommensurate with an undergraduate dissertation). This is unfortunate because relatively simple measures of central tendency – such as the maximum, minimum, mean, median and mode – or devices, such as the (nature of the) frequency distribution, can reveal much about the phenomenon captured by the variable.

What is more, such measures and devices can be useful in establishing the extent to which there are biases or skews in your sample (see above), the extent to which these may impact on your results, and whether your sample is a faithful one. For example, in studies of tourist behaviour it is common to find descriptive statistics of simple socio-demographic variables such as the age, gender, education or occupation, accompanied by a commentary on the extent to which they may be used as explanatory variables later in the analysis.

Many students employ bivariate statistical techniques to draw a series of inferences about the relationships and associations among variables within their datasets. In terms of selection, a second important point to recall is that the choice of procedure should be driven by the scale of measurement you have employed (as well as your intended outcome). For example, if you wish to examine whether there is a correlation between the disposable income of tourists and the length of vacation, you could use either the Pearson or Spearman's rank correlation depending on how you measured both variables. These are the parametric and non-parametric procedures for correlation respectively (Field 2009).

Finally, with respect to appropriateness you should recall that there are important qualification criteria that determine whether or not it is valid to not only employ a particular procedure *but to also employ its output*: to ignore these criteria means increasing the risk that your data are spurious. For instance, one of the most frequently used procedures in dissertations is the 'chi-square test of association' (note: it doesn't establish a relationship). Many a time we have been confronted by complex contingency tables, and output that suggests that the expected value in too many cells exceeds five or is less than one; that is, the output does not conform to the criteria for the test (Field 2009). Some students do not understand this, others forget to look for these data, and other choose not to apply the criteria. Whatever the reason/s, the likely outcome is spurious output.

Back to basics: reporting simple descriptives

Measures of central tendency are representations of your data. They tell the 'story' of your data in raw format. Some students, however, make the mistake of reporting either far too much or, as we have noted above, far too little of these data, thus resulting in a lack of clarity in the storyline. However, you should not dismiss simple descriptives altogether, but rather find an appropriate balance. In many instances, it is basic metrics (i.e. means, medians, modes or ranges) that deliver the most memorable moments for the reader.

Table 12.1 shows perhaps the most common way of presenting simple descriptives. There are several things that you should note here. First, only the proportions (or

Table 12.1 An example of tabulation for data reporting from a hypothetical dataset

What visitors do at the attraction:	Never (%)	Rarely (%)	Sometimes (%)	Always (%)
Visit the interpretation centre	—	8.5	74.0	17.5
Visit the café	1.5	9.0	70.0	19.5
Buy a souvenir	2.5	31.5	62.5	3.5

Source: authors

percentages) are reported. We often see many students reporting both the 'n' value (representing the total number of respondents who answered the question) in addition to the proportion. Doing so is an example of over-reporting as it provides the same data twice and thus violates the principles of simplicity in reporting.

Second, note that the proportions are reported here to the first decimal place. You should check your institutional conventions and, while you may wish to round up or down as necessary, the number of decimal places you use should be both appropriate to the phenomenon you are describing and consistently used through your dissertation.

Third, note that proportional sums are not presented. In other words, we have not indicated where a sum to 100 percent is possible. In some cases, this is appropriate, but quite often it is not necessary and results in more information being depicted than is necessary.

As a final point to consider, you may find yourself reporting quite a few simple descriptives in quick succession. 'Reader fatigue' is a real risk (see Box 12.3). Here the reader is literally being bombarded with one 'fact' after another until it is difficult to follow the point. One way of mitigating against 'reader fatigue' (which at the same time can actually

---

Box 12.3 'Reader fatigue' in the reporting on holidaymakers' preferences for visits to the UK

The inbound holidaymakers' travel preferences were examined by their choice of major attractions and activities in the UK. Overall, there was no significant difference in their preference for visiting England and Wales. Independent travellers were more likely to visit Scotland and Northern Ireland. In contrast, people on a package tour preferred Wales. Nineteen major sightseeing locations were listed in the questionnaire, and six attractions in London accounted for a total of 44.6% of the respondents' noted preferences. Outside London, Cambridge (10.1%) was the most favoured place to visit, followed by Edinburgh (8.6%), Oxford (8.6%), the Lake District (6.4%), and other places mentioned were: Manchester (1.6%), Liverpool (1.6%), Stratford-upon-Avon (Shakespeare's birthplace, 2.5%), Stonehenge (4.8%), Bath (3.8%), Cardiff (1.8%), Wimbledon (2.4%) and Windsor (4.3%) . . .

Source: authors

increase the readability of your work) is to find alternative ways of presenting your results. For example:

- *Original:*   65 percent of respondents planned on visiting the war memorial.
- *Alternative:* Slightly more than two thirds of respondents planned on visiting the war memorial.

- *Original:*   73 percent indicated that they read the guide book.
- *Alternative:* Just over seven out of ten surveyed read the guide book.

- *Original:*   Only 20 percent said they would 'definitely' visit the island again.
- *Alternative:* Only one in five visitors indicated they would 'definitely' visit the island again.

Reporting inferential statistical tests

Reporting the results of inferential statistical tests is not hugely different from simple descriptive statistics. The main difference, however, is the level of detail required. For most inferential tests, several values are provided and, to the trained eye, these indices reveal both the test you have conducted and sometimes something about the relationships you have unearthed within your dataset. For example, a Mann Whitney U test is a non-parametric test for differences among two samples (Field 2009). It offers a simple U value (a test statistic), but it is also important to report the p value (the level of confidence or 'significance level') of the result as well.

Simple tables are often used to report the results of multiple statistical tests, with enough columns as required to accurately and completely showcase the results. Often we find that students take the lazy option and cut-and-paste output directly from their software. This means that superfluous data that do not help make the statistical case are entered into tables, almost as a default position. For example, it is common to find expected values and cumulative frequency counts included in output generated from cross-tabulations for the chi-square test. These are not important in making the case as to whether an association exists or not, and whether a contingency table is valid or not (Field 2009). Hence, for an examiner, the inclusion of this output suggests that the student isn't confident with statistics because s/he is unable to discern which data to enter.

Of course, we have already warned you against duplicating the content of tables as part of your text. However, by providing a selective narrative you can demonstrate a greater level of critical engagement with your results. This can be done, for instance, by commenting on what you consider to be the important and unimportant features and your reasons for drawing such inferences. Your judgements as to the salient features demonstrate the extent of your grasp of the meaning of the output, particularly as it relates to your study.

For example, suppose a series of correlations were run on some data with varying results. Correlation coefficients range from -1 to $+1$, so it would not be unusual to see explicit values of, say, 0.4, -0.7, 0.9, and so on. When reporting correlation values (and the R^2, or coefficient of determination), you may want to make sense of the diverse values by applying some sort of order based on the effect size. As Field (2009) describes, standard prescriptions are that r = 0.10 is a small effect, r = 0.30 a medium effect and r = 0.50 a large effect. The interpretation of statistical tests is largely non-universal, unlike the process and mathematics of the test itself. These orthodox thresholds for effect size are contested (Aguinis and Harden 2009; Field 2009; Lance 2011). Thus, you should be prepared to justify why, for example, you deem a correlation coefficient you generate as 'mildly strong' or even 'weak'.

Running tests of significance in statistical software packages is simple and easy, but the danger is the misreporting of important information. While some of this should have been covered in any statistics courses you may have taken, there are a few situations that deserve clarification before you start writing up (Chapter 13).

First, anytime you are discussing results of statistical tests within the text of your dissertation, you should immediately provide the statistical test results. For example: 'The results indicated that male tourists were more likely to participate in hang gliding than female tourists ($U = 0.567, p = 0.03$).'

This indicates that a Mann Whitney U test was performed (hence the U value) with the p value shown to establish the level of the significance. The way this was written is clear for reader: there can be little doubt as to what statistical test was performed, and that it was legitimate to accept (the alternative hypothesis) that there was a statistically significant difference at the 95 percent level of confidence.

The interpretation of p values can be just as subjective as the interpretation of correlation values (Gelman and Stern 2006; Field 2009; Aguinis et al 2010). The trick with p values, however, is that there are often differing values depending on the confidence interval selected and your willingness to commit a so-called 'Type I' error – when you believe there is an effect but there isn't (Field 2009). Put another way, this is when you reject the null hypothesis in favour of the alternative hypothesis (i.e. there is a difference) although it is the null hypothesis that is in fact true. For example, a p value of 0.06 is not significant at the 95 percent confidence interval, but it is at the 90 percent confidence interval. By reducing the confidence interval and accepting the alternative hypothesis you are increasing the chances of drawing a false conclusion of an effect or difference.

While we have concentrated on difference among these examples, it is important to remind you that 'null results' (i.e. those that demonstrate no difference) should not be casually dismissed. While you may have expected, possibly even desired, differences to be revealed (perhaps based on precedents in the literature), the fact that they were not observed in your study is both a valid and worthwhile observation. Some students are tempted to discount a lack of observed difference as in some way pointless when in fact in some circumstances it can be quite meaningful. For instance, you may have reason to believe, based on your literature review, that there will be gender differences in the propensity to engage in adventurous activities while on holiday to New Zealand. Were your survey to demonstrate no significant difference, this would a meaningful result.

Finally in this context, you should not disregard 'ambiguity'. Many questionnaires employ Likert scales based on directed statement and these frequently include (usually) a middle positioned statement such 'neither agree nor disagree'. Moreover, they may also include the facility for respondents to enter that they simply 'don't know' (or there may be a facility if they do not have/want to enter an opinion). While the intricacies of these design features have been extensively discussed elsewhere and they are long-standing issues (Ryan 1995), in practice we find that many students readily overlook these two important aspects of their data. In the case of the latter, 'don't know' might be a function of a poorly drafted question and/or a complex idea. Hence it may tell you something important about the mechanics and experience of completing your survey from a quality assurance perspective. However, it may also indicate an important lack of awareness about a key issue or that a particular (complex) issue, such as climate change or sustainable behaviour, is not well understood among your participants. In the case of the former, students routinely look for polarized views by conflating values for agreement and disagreement. However, a modal value of 'neither agree nor disagree' may be notable because it may indicate that respondents really have not been able to decide one way or the other about an issue and instead ambivalence is the order of the day.

Some final advice on reporting quantitative data

If we were to develop some principles about the reporting of quantitative data, they may look something like this:

- *Rule 1* – never assume that the reader knows (or remembers) what kind of tests you have performed.
  Even if you discussed it in your methods chapter, make it very clear. Often this can be accomplished by including procedure-specific information in the reporting of the statistical test (such as the Mann Whitney U test example above), but other times it might be necessary to indicate overtly what test was conducted.

- *Rule 2* – avoid 'over-reporting' information that is not necessary.
  In other words, always report only the most critical information from a statistical procedure. For example, in some situations, reporting the upper- and lower-bound values on standard deviations might be important, but for the most part you can simply report that absolute value of the standard deviation.

- *Rule 3* – Greater variety enhances readability.
  An appropriate mix of tables, graphics and text represents a more effective form of communication and it is likely to hold the interest of readers for longer.

- *Rule 4* – Never copy and paste the raw output of a statistical software package directly into the text of your dissertation.
  The reason should be obvious, but in case it is not, it is because the layout of the output was never designed to feature in a formal text document such as a dissertation. Often it contains much more information than is necessary for reporting of your results.

Qualitative reporting

Several of the general ideas – such as the need to analyze with a purpose and to communicate your results effectively – are valid in all contexts. Nevertheless, qualitative data, by its nature, is generally much richer than quantitative data and its analysis tends to be a more iterative process that involves the reading and rereading of texts to make sense of them and what the data reveal about your topic.

Specific techniques in qualitative analysis can be found in a range of generic methods texts as well as those dedicated to their use in such disciplines such as anthropology, human geography, social work and sociology. Indeed, if your research utilizes qualitative methods, it would be wise to familiarize yourself with the latest methods and practices of analyzing qualitative data for two reasons. First, as Silverman's (2011) collection indicates, there is rapid innovation in qualitative research and, as a result, it makes sense to remain up to date. Second and connected, there is a much wider range of techniques and approaches to analyzing qualitative data. For instance, Creswell (2012) identifies what he terms five broad approaches – originally five traditions (Creswell 1998) – among a multitude of methods: narrative research (i.e. earlier 'biography'), phenomenology, grounded theory, ethnography and the case study. Rapley (2011) notes that many novices are in a quandary about which of the analytical approaches to use and identifies four of the more routinely cited: framework analysis, thematic analysis, interpretative phenomeno-logical analysis, and constructivist ground theory.

What follows, therefore, is a brief overview and synopsis of what is possible in qualitative analysis. Just as gardening requires you to get your hands dirty, so too does qualitative analysis, metaphorically speaking. It demands a thorough understanding and review of both the techniques you are proposing to employ as well as the data you have collected. You have to be patient, and it can be very time consuming (but highly rewarding). But first, a few practicalities:

- Your data from qualitative research are often the written word (i.e. text). This is generally in form of transcripts from interviews which were recorded, but also so-called 'field notes' that you may write as you conduct interviews. The latter can be very useful in providing context for interviews, and you should not omit to make notes during the interview about how people respond to the questions or the experience of being interviewed because this too can be important context.
- As a general rule of thumb, if you record an interview (see Chapter 9 for an overview of the ethics of such a procedure) the entire interview should be transcribed. As time consuming as this can be (see Chapter 7), resist the temptation to just pull out quotes as you listen, otherwise you may miss the all-important context of a particular quote or concept as elicited by your respondent. A more selective approach is also risky if you think that your writing up may be punctuated by intermissions.
- For practical purposes, retain one set of master transcripts which are never touched, just in case you need to revise your analysis through an entirely fresh reading.

Coding qualitative data

We start by reviewing what constitutes qualitative data. Quantitative data are decidedly numeric, and thus their treatment is arguably more straightforward. In contrast, qualitative data can take many forms, including interview transcripts, notes from observational techniques, visual images, sounds, focus-group transcripts and even answers to open-ended questions from quantitative surveys. Ultimately, this raw data gets translated into text as you try to make sense of it and jot down your ideas, notes and make records. Then it is your job to establish, efficiently and convincingly, context and meaning from within that prose.

There is no single correct way in which to code qualitative data (Bryman 2004) but coding is about classification and being able to move between sources and their authors to examine what is present and absent about particular issues or on specific subjects. Many students start with a print copy of, for example, an interview transcript, and they systematically identify broad categories (or even wider themes comprised of several categories) as they read through the first and then subsequent transcripts, refining their ideas along the way. Such an iterative process forms the essence of coding, and it is one that can actually yield excellent results despite sounding 'low-tech' compared to the analysis of quantitative data using powerful software. Coding involves a review of, for example, interview transcripts to look for illustrative or benchmark issues (which may become categories) and any associated words or phrases. It can also mean taking other forms of qualitative data (e.g. reports or online data such as discussion forums) and deconstructing those data into smaller, more meaningful 'bits'.

As you might expect, coding requires careful attention to detail, and we can offer a few practical words of advice in this regard:

- Keep a list on a separate page (or in a separate file) of all of your categories. As you identify additional categories, this can help prevent duplication. It can also function as a means of identifying larger themes from a collection of similar categories.
- A more advanced technique might utilize longer codes that become associated with a particular quote, word or phrase in your transcript. For example, ANGER-233-1-34 might refer to the category of 'anger' on page 233, with '1' representing the first quote, word or phrase associated with this category and '34' representing the respondent number. Eventually, you may be able to cross-reference quotes, words and phrases more easily using such a method.
- While some (e.g. Charmaz 2004) advocate assigning at least one code for each line of an interview transcript or secondary document on which you are conducting analysis, this means that you may end up with hundreds of codes which can be difficult to string together effectively in order to parse out meaning from the data. We would recommend that you start by evaluating your data carefully and looking for major categories first. From there, you may be able to identify sub-categories.

In the end, you will generate several categories of issues and several associated quotes that you can use to illustrate respondents' perspectives. Note that coding also works for other forms of qualitative methods, notably notes obtained from focus groups, observation sessions or the content analysis of secondary documents.

Some final advice on reporting qualitative data

While there are many approaches that may be used to report qualitative data (Creswell 2006; Sandelowski and Barroso 2006; Corbin and Strauss 2008) and software packages potentially to assist you such as NVIVO (Bazeley 2007) or ATLAS.ti, there are several basic things for you to recall as you conduct your analysis and subsequent writing up:

- *Be analytical!*

This is not as axiomatic as it may appear. Students often forget to consider the basics. Qualitative research is not a 'soft option', nor should it be considered less scientific or rigorous than quantitative research. The range and level of issues raised as well as the degree of consensus and dissonance among respondents are important. Look for themes and trends within your data as well as patterns and variations, perhaps based on simple socio-demographic differences among your participants.

- *Weave together your narrative*

The 'writing up' of qualitative data requires more 'flow' in the text than quantitative data. This is largely because raw qualitative data is often presented directly within the dissertation, and thus the text must be shaped in order to maintain the flow of the argument and its structure. For this reason, you should be quite careful about how the direct, raw data you use in your text fits with what you are saying. Quite often we encounter snippets or quotes in qualitative data analysis that do not add to the present discussion. What the student has done is simply included a quote from an interview for the sake of including it.

- *Silence can be golden*

Remember, silences and what is not said are often as revealing as what is said. Don't be afraid to comment on them or on the style of the response. For example, a poor, uninformed or inaccurate answer to a question may reveal something important about your respondent/s.

- *Contextualization*

Some students present qualitative data derived especially from interviews and focus groups in a decontextualized manner; that is, without reference to the background of the respondents. In doing so, they overlook one of the main premises of qualitative research, namely that behaviour (i.e. opinions) are a function of the positionality of the participant. There are various devices used for this purpose including thumbnail sketches and/or tables of characteristics of the participants.

- *Anonymization*

Your reporting has to comply with the ethical conventions of your institution as well as the ethical parameters you have set yourself (see Chapter 9). For instance, you must not use direct quotes from those respondents who have expressly requested this as a condition of their participation. Likewise, you may have to present a copy of the narrative or quotes to respondents for checking. Finally in this regard, be careful that, if you do present tables or thumbnail sketches of respondents as contextualization, there is no chance that respondents may be identified and hence views attributed to them.

One question many students ask is, 'How much data do I include, or how many "quotes" should I present?' Frustratingly, this is difficult to answer. There is no single standard or a magic number of quotes to include, but most examiners will know immediately if there are not enough or too many. When there are not enough, the text appears to be making generalizations or the 'voice of the respondent' is mute and conspicuously absent. When there is too much, the text almost reads as if your interview transcripts are being annotated. In practice, you may find that the constraints of word count weigh heavy in your final decision on how many to include.

Reporting on more than one type of data

Of course, you may have elected to employ more than one method in the data collection for your dissertation and this presents you with two distinct challenges for the analysis and presentation of your results.

Firstly, using more than one method requires you to conduct thorough and systematic analysis of the data generated by each method. This means that you should be fully conversant with, and able to apply, analytical techniques for each method. Put another way, this does *not* mean that if you are employing two methods for data collection you only need to be aware of half of the ideas associated with the analysis and presentation of each one! Partial and underdeveloped data analysis is the hallmark of weaker dissertations (Chapter 15). If you decide to employ more than one method, you have to commit to understanding each method, its principles and traditions, as well as best practice in the analysis and presentation of data.

Secondly, you must carefully consider how to present the data generated from the different methods, not least because you have a finite word count (Chapter 14) and you are not rewarded with extra capacity for writing up more than one method. As we have noted previously, there are differences between mixed and multiple method approaches. With respect to the former, the data generated by the respective methods are analyzed and presented separately, almost 'siloed' from one another. This may be because different methods address distinctive and different objectives. For instance, some objectives may be exclusively researched by interviews while others may only require data from a content analysis. With respect to the latter, careful consideration is required because a more integrated approach is required, involving a greater degree of synthesis. Your research will have been designed to be complementary and reinforcing in terms of data collection and your analysis should also reflect this. You may, for instance, have to think through such issues as the sequence of reporting, the relative importance you are going to afford the results generated from each type of method, and how the respective results function most effectively to deliver your objectives and answer your research questions.

Turning results into findings: four out of five visitors preferred . . .

So far you may have noticed that we have been careful in our use of vocabulary: we have referred to data and results, and we have avoided the use of the term 'findings'. This is because there is an important distinction between results and findings: in short, findings are results in context. You will generate many results from your analysis but you must indicate to the examiner which results are the most important (or more important than others) with respect to your objectives (and your research questions), and you must explain why you have afforded them such status.

This may also be because certain results are simply so compelling and emphatic in nature. In the wonderful world of advertising these are often referred to as 'killer facts', and we can all recall key pieces of data from TV and newspaper ads. Another connected term is 'take home messages': this refers to what message/s an advertiser wants the audience to take from the communication. While it may seem somewhat trivial to invoke these ideas in the context of an academic piece of work, what do you want your reader to recall or quote from your dissertation? This is one aspect of writing critically which is a hallmark of stronger texts. If you still need to be convinced, a key skill fostered by the dissertation process is the ability to review your own work (Chapter 14). Moreover, you have to decide the main outcomes if you are to write an effective conclusion (Chapter 13) and to encapsulate your research in a representative abstract (Chapter 14).

At first inspection, this advice may appear to have greater resonance with reporting on quantitative data analysis. After all, this can generate a large array of decontextualized data for you to make sense of. It may even be a useful distinction to make if you are employing content analysis in tourism research (Hall and Valentin 2005). However, it is important to note that in many types of qualitative research there is a necessary, if not always automatic, conflation of results and findings in reporting. This is because critical analysis cannot be divorced from the ideological commitment of the researcher, who is always and already predisposed to look for particular meanings and not others. Theory is always present in the act of analysis and within the results chapter/s there is a need to provide a continuous commentary that connects results and findings (usually in the results chapter). This may be especially the case if you employ critical discourse analysis or some forms of ethnographic research which are becoming widespread in tourism research.

With this distinction in mind, it is nevertheless worthwhile noting that – whatever approach you take – invariably the *relative* importance of your observations will relate to their contribution to addressing your objectives (and/or answering your research questions). Very often, as the previous paragraph implies, this is established by making comparisons between your research and external reference points. In quantitative research in particular, your observations may be compared with results and findings presented in the academic and/or grey literature (Chapter 4). After all, as part of your objectives you may have been attempting to use a particular method, examining a particular result, or testing a specific model in a different context of application. Alternatively, your results may cast previous findings in a different light or suggest the need for further follow-up research which is a key section within your conclusion (Chapter 13).

The chapter at a glance

The main learning points of this chapter are that:

- Data analysis within your dissertation is about the effective generation, interpretation *and* communication of results.
- In the course of your data analysis you should seek to establish the credibility and reliability of your results.
- You should carefully plan out your data analysis, ensuring that it addresses your objectives and answers any research questions.
- As much rigour is required for making sense of and reporting qualitative data as it is for quantitative data.

Dissertation checklist

Before you go further in your work, check you:

1.	Have revised your prior learning and training on analytical techniques before commencing the analysis for your dissertation.	
2.	Know where to obtain detailed guidance on the principles of, and best practices in, data analysis.	
3.	Have access to the necessary analytical software (for both qualitative and quantitative data) you are going to use for your analysis.	
4.	Are aware of the types of data and/or analytical techniques used in similar studies which may have informed your work.	
5.	Have taken the relevant results and valorized their importance as (key) findings.	

13

STRUCTURING YOUR DISSERTATION

Learning outcomes

By the end of this chapter you will be able to:

- Recognize the key components within each chapter of your dissertation.
- Understand the roles and internal functions of each chapter.
- Compile the main text of your dissertation.
- Signpost your reader through your dissertation.

Organizing your written work

At some point in the dissertation process you will need to develop an outline of your work. In its most basic form this is usually just a listing of the chapters, or sometimes a more detailed chapter-by-chapter breakdown by section and sub-section. In some instances you may be asked for a preliminary plan as part of your research proposal in order to establish what your dissertation may look like in the future. Commonly though, your advisor/s may ask to see it as you gather your ideas before starting to write up.

This chapter revisits two of the key ideas presented very early in the book, namely that: the dissertation is comprised of a series of discrete elements, each of which fulfils a particular role in your final text; and the importance of establishing connections running through your dissertation. Instead of disassembling your dissertation as we did in Chapter 2, the emphasis in this chapter is not only on how to put the various components together but also how to weave a strong thread through your entire text.

Both aspects are of critical importance, but in many cases insufficient attention is paid to establishing such linkages which is why we cover them both here and in Chapter 14. Here, the focus is on establishing the connections through the arguments you develop while the next chapter shifts to enhancing internal continuity by more effective presentation and self-appraisal of your document. Thus, taking care over your structure

---

Box 13.1 The possible structure for your dissertation

Title page
Abstract
Table of contents
Lists of tables
List of figures
List of appendices
Introduction
Literature review
Methods
Results*
Conclusion
List of references
Appendices

(* There may be more than one, also called 'analysis' in some programmes (see Box 13.4))

Source: authors

---

is a vital first step towards producing a good dissertation and is an important part of the research process.

There are conflicting ideas on how dissertations should be structured. It is rare to find two dissertations that have precisely the same structure when the composition of sections and sub-sections is considered. Beyond this micro-level view, there is no need for us to cover all of the possible permutations of potential dissertation structure. In what follows we present to you a blueprint that many of our students have adapted with success to the particularities of their own research circumstances (and which was previewed in Chapter 2). This basic structure in Box 13.1 – which eschews the use of separate chapters for background and discussion – nonetheless includes the key components required of most, if not all, dissertations in tourism studies (and in the social sciences more generally).

In addition, it is worth noting one other very common feature regarding the chapter headings. Compare their simplicity to the more descriptive and elaborate titles used in many books for example, this one included. The best way of understanding this structure is to go through each chapter in turn to remind you of its purpose, and to describe in more detail its internal structure and the sort of content it should contain.

Chapter 1: Introduction

At this stage, it is probably worth recalling the idea that a dissertation should function like a novel in so far as you tell them – your readers – what you are going to tell them, tell them what you are telling them, and tell them what you told them (see Chapter 2). As a result, a clear and compelling introduction is needed so that, by the end of it, your reader is absolutely clear about your subject, why it is an appropriate choice, and what is about to unfold in terms of the nature and progression of the arguments. Accordingly, there are normally three main parts to this chapter, namely:

1. *Research context*
This is used to set the scene and demonstrates the importance of the topic. In addition to a brief academic rationale (or justification) which summarizes the *main* intellectual arguments for conducting the research (i.e. a *preview* of the literature review where they are made more extensively), this chapter establish the relevance of the research in its economic, social, cultural, environmental and/or political contexts. In effect, this section of the introduction also provides key (non-academic) background to the study (if a separate background chapter in its own right is not permitted and/or justified).

2. *Research aim/s and objectives*
Within the introduction the main strands of continuity are set out for the first time. It is not necessarily important to elaborate the full range of research questions or hypotheses (if you have them) at this stage (these might be elaborated in the methods chapter). Rather, it is important to establish, in basic terms, what you intend to do and why this is justified. Your notes from Chapter 3 (here) might assist in this regard.

3. *Structure of the dissertation*
You should give your readers a (general) preview of the chapters to follow by providing a précis of their content (potentially a paragraph on each). This does not mean telling them the specifics or details of the major points you are about to develop or offering key results or findings early on. Rather, it is about explaining how the dissertation is structured and what role/s each chapter *will* serve in order to deliver on the aim/s and objectives.

In light of the specific content of the introduction, it is perhaps a lot clearer now why your introduction may be one of the last chapters you write.

Nevertheless, taking our simple metaphor of the 'dissertation as novel' forward, at the end of the introduction you've told the reader what you are going to tell them. In the next chapter, it is time, as it were, to start telling them what you want to tell them.

Chapter 2: Literature review

This chapter is where you demonstrate your knowledge of the literature, but remember you should do so in terms of your aim/s and objectives. The idea of writing with a purpose becomes critical in the literature review. In this context you need to ensure that you write in a less descriptive way and attempt to develop what can be called a more critical writing style.

Critical writing was introduced in Chapter 4 where we also examined some of the basic features of what makes a good literature review. This is because some programmes require dissertation students to submit (initial) literature reviews as part of the proposal. We have also encouraged you to be critical in your choice of approaches, methods and techniques by making informed, properly reasoned decisions on their relative merits (Chapters 5, 11 and 12). It is nevertheless worth returning to the idea of critical writing just briefly.

Many students struggle with what this means in practice. Let's be clear: being critical isn't about the annihilation of other people's work or taking an overtly negative or cynical view that only brings out shortcomings or limitations. It is about making evidence-based arguments that logically lead to appropriate conclusions. Specifically, within the literature review, there are five basic ways in which you might write critically (see Table 13.1). With respect to the analysis of an idea's relative merits, Ridley (2008) notes that this can result in two possible outcomes, namely that: you will eventually agree, confirm or defend a position on the basis of the relative merits; or you may have to concede that a position has some strengths but that there are also certain important weaknesses. Put another way, when you are asked to assess a set of ideas it is insufficient just to draw attention to their strengths (or weaknesses). You have to look at the balance. As the table indicates, you can also reject ideas from the literature as long as you have sound reasons for doing so.

To illustrate this Box 13.2 shows a hypothetical example of a problematic literature review from a student dissertation. This has no references to support any of the statements and no clear introduction. In fact, it has no engagement whatsoever with the academic literature in this topic area and it might just as easily be described as background. Indeed, this is often one of the problems characterizing work by weaker dissertation students: namely, they misunderstand the literature-review chapter and its purpose, describing the 'grey literature' at the expense of the academic literature. Were it to be have referenced, the chances are that many of the figures supplied here might have been sourced from reports. As it stands, a strong case could be made that, in its serial absence of appropriate citation, it plagiarizes (see Chapter 4). Remember, basic ignorance is not mitigation.

Table 13.1 Writing critically in the literature review

Action	Purpose
Comparing and contrasting theories and concepts	To indicate the position you will take in your dissertation
Strategic and selective referencing	To demonstrate your command of the literature on your topic and to underpin the arguments on which your dissertation is based
Synthesizing and reformulating arguments from one or more sources	To create new or more developed perspectives
Analyzing an idea's relative merits	To defend or contest on the basis of the relative weight of evidence in favour and against
Explaining the reasons for rejection	To reject a position using proper evidence such as inadequacy, lack of evidence or fallacies

Source: adapted from Ridley (2008: 119)

Box 13.2 A hypothetical example of a problematic literature review

China's outbound tourism has become stronger and grown faster than ever before. The following figures show huge increases. Figure 4 shows China's outbound departures and annual growth. As can be seen, growth peaked in 2004 with a 42.69 percent increase compared with the year before. More than this, the figure was increased to 31 million in 2005 and estimated to increase further in 2007. As Chinese citizens' purchasing power is continuing to grow and demand has increased for international travel, the consumption of tourism products has continually expanded during the last decade despite the impact of SARS in 2003 . . .

Source: authors

In contrast, Box 13.3 illustrates an example of a more promising discussion that looks at different perspectives and supports its ideas with strong referencing. Here the student has skillfully blended several academic and practitioner sources to make the points that the disabled traveller's needs are equally as valid as the non-disabled, but that the former's needs are routinely and indeed systemically neglected. The selective use of citation makes for a more interesting read, while the 'real world' relevance of the research is demonstrated in addition to the academic originality and contribution it will make by filling an apparent research gap.

In terms of structuring the chapter it is essential that you provide a clear framework, usually based around key approaches or perspectives you have identified within the body of knowledge. This may involve a more advanced technique known as 'meta-analysis' which we introduced in Chapter 4.

Meta-analysis is becoming increasingly popular at all levels of research. It provides you with the opportunity to develop a solid evidence base against which to justify your conclusions about the extant body of knowledge as it relates to your research objectives. If used as an organizational device, your meta-analysis table will help you provide a structure along with your research objectives. The main parts of this chapter will be those key research themes identified in the meta-analysis and/or from your reading.

You can develop and hence employ both thematically related (i.e. ideas, concepts and theories) and/or methodologically related (i.e. approaches, methods, techniques) meta-analysis tables. These could help you to structure your literature-review chapter and/or your methods chapter respectively. You do not have to place a methodologically related meta-analysis table only in the methods chapter. This is because one of the key aspects to come from your review of the literature is that there are varying and contested approaches to the production of knowledge in your subject and how your subject is known and understood. Epistemology and ontology, respectively, are at the heart of a literature review and they are manifest in the range of methods used (Gallarza et al 2002).

Box 13.3 A more promising literature review

Holiday selection is typically longer and more convoluted for the disabled traveller (Darcy 1998; Veitch and Shaw 2004). Information deficit, validity and accessibility are all challenges faced by the disabled potential tourist. This is not a simple matter to address. As Holloway and Robinson (1995: 163) comment on legislation to ensure brochure accuracy:

> To illustrate how. . . . this can work against the interest of the consumer, facilities at a hotel which could be of help to disabled travellers may be ignored rather than risk the possibility that they were inadequate: for instance, 'wheelchair accessible' could be true for the majority of wheelchairs but can the operator be certain it is true of all makes?

Yet, so-called 'honesty in brochures' to borrow from Swarbrooke (2003: 77) applies just as much to disabled consumers as to non-disabled consumers. Honesty and validity are basic consumer rights, and they depend on needs and wants. In this regard, the informational barriers to tourism have been little researched beyond the acknowledgement of needs for information access and the meeting of legislative demands. A survey by VisitBritain (2003) highlighted the discrepancies in information often faced by people with disabilities such as hollow claims of disabled facilities and reassurance of suitability. Such experiences generate a need for an 'informational guarantee' but how this might work has not been the subject of research to date . . .

Source: adapted from Salt (2004)

For this chapter, and indeed all your (subsequent) chapters, it is important to have a short introduction that gives it a (brief) context and structure. Remember, a good introduction to each chapter should cover its:

1. *Aim* – this should highlight the purpose of the chapter (within the context of the dissertation as a whole); and
2. *Outline* – this should give a very short introduction to each following section.

These short introductions set the scene and they provide your readers (i.e. examiners) with a 'roadmap' setting out the milestones through the text. To be clear, you can use a 'roadmap' for the entire dissertation (i.e. the third part of the introduction) at a macro-level as well as at a micro-level within each chapter. Here you are using the roadmap to point the reader towards what you are about to say in substance about the literature you have been reviewing as part of your dissertation.

Finally, as one of the main chapters of your dissertation, it is useful to remind the reader what you have told them. After all, you may have covered quite a lot of ground.

Remember, if you are intending having a concluding section to the chapter, it is effectively a staging post for the reader. It should not just be a general summary. It should draw together the key ideas in the context of your objectives and suggest the implications of the recently read content moving forward. Write with a purpose!

Chapter 3: Methods

This is the chapter that allows you to demonstrate your knowledge of the principles and practices of empirical research. A good methods chapter leaves no doubt that the research was properly planned, executed and reported. It is also important to note that methods chapters should include information on the analytical techniques as well as the methods of data collection. While the latter might be the subject of more attention in the chapter, discussion of the former is certainly necessary and should not be overlooked.

As we mentioned earlier, the methods chapter is an important bridging point between the two 'halves' of your dissertation (Chapter 2); that is, from setting out the case in the introduction and literature review in the first 'half' to conducting and reporting on your research in the second 'half'. The methods chapter serves two functions, enabling you to:

1. *Defend the methodological decisions and choices you made in order to conduct your research.*
There are many possible methods and techniques available to you, so one important to set of tasks is to appraise the various options that were at your disposal as they relate to your: aim/s and objectives; prior (analogue) studies within the academic literature; best (and good) practice in the generic body of knowledge on research methods; and their practical application in relation to your research.

2. *Explain how you executed the empirical work in terms of both data collection and analysis.*
Possibly more functional in nature, you should explain in some detail issues such as when, where, why and how you conducted your work as well as any obstacles you encountered along the way and how you overcame them. Taken together this information provides the foundation from which you are able to assess the quantity and, more importantly, the quality of the data you have generated. It also provides the means by which subsequent researchers could follow your method if one of their objectives were to corroborate your work or to replicate it in another setting.

In both cases, the ultimate goal is to establish the extent to which you have been able to generate valid, reliable and trustworthy data that allow your dissertation to address your aim/s and objectives. In other words, you are establishing whether readers can have confidence in the results you report and the conclusions you draw at the end of your dissertation.

There are limitations associated with all forms of method (and analytical technique), so an important aspect of the methods chapter is to be as open, honest and transparent about what you have done as possible. This means articulating the shortcomings in your approach as well as, more predictably, extolling the (no doubt, many) virtues and strengths of the decisions you have taken and methods you have employed. The ability to reflect on your decisions and to assess the relative merits of your work are hallmarks of greater competence in research.

A common mistake students make is to spend too much time discussing the broader aspects of methodological approaches at the expense of explaining exactly what they did in their research. Put another way, principle often wins out over practice. The key thing is to have a balance between these two fundamental aspects, and this is achieved by following a clear structure.

Structure of the methods chapter explained

Once again, this chapter requires an introduction to signpost the reader through the progression of your ideas.

It also needs to be set out in a clear fashion. Following this you can provide a general discussion of the characteristics of various methodological approaches. In this discussion you need to highlight the potential advantages and disadvantages of different methods relative to your research objectives. It may appear somewhat counter intuitive, but sometimes, in order to justify what you have done, you have to demonstrate that alternative methods were not fit for the purpose of your research.

Based on a review of the relative merits of various methods to deliver your objectives, you explain which tools or instruments you chose as most appropriate to your study. It is here that the table relating objectives to data required and survey methods might be located (see Chapter 11). This could provide a strong focal point for your discussion.

You must justify your methodological approaches, not just describe them. The major decision is usually whether to use quantitative and/or qualitative methods. One other way of defending which method/s are suitable is to cite previously published research on your topic. This can be taken on from your literature review (Chapter 4), even drawing on the content of a meta-analysis table (see above and Table 4.3).

This section is important because it sets your survey methods in a wider context, it demonstrates that you understand the different approaches available, and it shows that you have the ability to be critical in your decision making, which is a hallmark of stronger dissertations. Having established the range of possible methods, you now need to move on to the details of the data-collection methods *you* actually used. This is where you need to explain clearly exactly what you did.

Key details of how you generated your data and results

Initially, you need to cover at least *two* key aspects:

1.   The development of your survey instruments; and
2.   How you generated your sample.

Included within the latter, you should provide a discussion of the sampling frame (i.e. where you chose to conduct the research and the justification/s for the choice of location/s) and the period of research (i.e. the start and finish dates). This is because your results may to some extent be a function of the time at which the research was conducted, such as in 'peak season', 'low season' or the 'shoulder months'.

Let us assume that you have decided to conduct a questionnaire survey. In addition to design and piloting, you will also need to explain how it was conducted and any issues that arose during the main data collection episode that could not be anticipated by any pre-testing.

Whatever the method/s you used, in this part of your methods chapter you must explain in detail:

- Your survey instrument/s in terms of design, content and pre-testing.
- The sampling strategy you adopted with each method.
- How you contacted your sample/s.
- How the data-collection episode/s actually proceeded.
- Conditions 'in the field' when you collected the data.
- Ethical (and other practical) issues you encountered and how you overcame them.

It is usual to put copies of your survey instruments such as questionnaires or interview schedules in appendices at the back of your dissertation (see Box 13.1). You can put in early 'piloted' drafts as well as the final drafts in order to demonstrate the due diligence you have observed and the evolution of the survey instrument/s.

If you have used multiple or mixed methods, you should explain how the methods were employed together as a complete package. For instance, it is common to find questionnaire surveys deployed with semi-structured interviews so it is important to explain here when and why the methods were used, because interviews can be used to:

- Identify a series of issues to cover within a questionnaire survey; and/or
- Examine issues raised from the analysis of a questionnaire survey in greater depth.

So far we have been discussing collecting primary data, but in some cases you may be much more reliant on secondary data. In this case your discussion here should highlight the accuracy and reliability of this information, and if possible discuss how it was collected and – if it is a sample – what the sample represents. This is just as significant for secondary data as it is with primary material. Secondary data are not immune to biases and skews which were introduced when they were originally generated. Such a discussion will demonstrate that you are fully aware of the need to consider the probity of secondary data as a major determinant of their trustworthiness as a foundation for your subsequent analysis.

The last part of this chapter concerns the analytical technique/s used on your data (see Chapter 12). In this part of the account, you are likely to set out how the raw data were processed, including details on the data entry, as well as data cleansing that was done, and which verification procedures were followed (to establish validity and reliability). You should also detail how the analysis was undertaken. In the case of quantitative material, your reader (i.e. examiner) will undoubtedly want to know what kinds of statistical tests you have conducted and the reasons for them. Rather than wait until the results chapter where it may become (more) obvious, it helps to discuss the tests themselves and relate their use (in theory and practice) back to the literature. Qualitative research is no different. In light of the array of techniques that can be used to process and interpret such data (Chapter 12), the specific procedures you used should be set out clearly for your readers.

The analytical procedures you used should have been appropriate to both the objectives for your study as well as the type of data you have generated. For instance, you may wish to explain why you chose to use a Kruskal Wallis test for analysis of variance rather than its parametric equivalent, ANOVA, as well as the implications of this decision (in terms of post-hoc tests).

Results: Structuring your results chapter/s

The number of results chapters you have depends to a large extent on your word limit, the number of your research objectives, the approach/es you have taken in your research (i.e. reporting quantitative and qualitative results separately) and/or your institutions regulations. You may only be allowed to have a single results chapter!

In one sense these chapters tend to follow a similar structure. Once again, you need a good introduction that gives the purpose of the chapter/s and signposts the reader through its/their structure/s. It is important in the introduction that you highlight which of your objective/s are being discussed in each chapter.

When students use a quantitative approach via a questionnaire survey, they commonly start by describing their dataset. This description of the sample (possibly against the background population) will establish the extent to which there is any bias or skew to the results. This discussion – which is essentially a data quality assurance exercise – can alternatively be entered in the methods chapter if regulations dictate this, word counts require it, and/or to ensure that the results chapter is exclusively devoted to an analysis of the dataset as it relates to the objectives.

If it is located in a results chapter, such a discussion is followed by an analysis of the data in terms of your research objectives and based on your tabulation plan if you used this device (Chapter 12). Often we see students progressively adding layers of complexity by starting with univariate analyses, progressing to bivariate analyses, then to multi-variate statistics if they have been used. This presents a wealth of statistical material for the reader. Of course, this is perfectly acceptable as long as it allows specific objectives to be addressed. This is key, and as a result some students prefer to structure their results chapters into sections based on objectives (see Box 13.4). Alternatively, the analysis can also be structured around answering specific research questions or addressing particular sets of hypotheses if you have elected to use these as structural elements.

A similar sort of approach can be employed for groups of respondents who have partici-pated in qualitative research. Based on an examination of a series of contextual socio-demographic characteristics (perhaps presented in a table presenting thumbnail sketches), you may identify certain features among your respondents that may position or even skew your analysis. This is normally followed by an examination of the categories or themes to have emerged from a critical reading of the narratives, subsequently structured around any patterns based on certain apparently explanatory variables.

There are three key points to recall in your reporting:

1. *You should report with a purpose, in a structured and directed manner.*
There is little point, as some students do, in presenting literally every last piece of data they can within the word count available. This is the reporting equivalent of 'carpet bombing' your reader with information in the vague hope that some will hit the desired mark and be relevant to their topic. Not surprisingly, this is usually a characteristic of a weaker dissertation.

2. *You must not just describe your results but rather write in a critical way.*
This is not an easy thing to do but it is important. Not every piece of data or result that you generate is equally important to your study. Likewise, not every objective has to have an equal treatment in terms of the volume of data presented or the number of key results you identify. You should indicate to the reader which results you think are more or most important and explain why. This may be because they are remarkable or unexpected. The importance of particular findings is revealed by comparison to data in previously published material. Don't make your readers guess or make the connections for themselves.

One way of getting a feel for this way of writing is to read a journal paper you find interesting and note how the author deals with the results of their analysis. This type of writing provides explanation as well as description, it should be focused on your research objectives, and it should be appropriate to which pillar/s of research is underpinning your research (Chapter 4).

3. *The style of reporting for mixed methods studies should be carefully considered* and it should reflect a combination of the sequence in which the work was conducted and the purpose of mixing. If you are using multiple methods to address particular objectives or research questions, you should attempt to triangulate the data you have generated.

Your concluding chapter

The time has come to recap and, in keeping with the metaphor, to tell the reader what you told them. The final chapter of your dissertation has, then, a well-defined structure which is as follows:

* *Introduction* – which in this case should be short and mainly focusing on signposting the chapter.
* *Summary of main findings* – this section may be simply structured around each of your objectives so you can demonstrate to your examiner that you have achieved what you set out to do in the introduction. As space may be at a premium, you will not be able to rehearse, once again, all of the findings from the previous chapter. This is perhaps no bad thing because it will force you – in a form of critical filtering – to decide upon which are the *most* important findings to draw to the reader's attention.

- *Key contribution/s* – often this is not required or used in undergraduate dissertations (but is certainly increasingly expected these days in theses at masters and doctoral level). This highlights the key contribution/s of your research to the wider academic debate.
- *Limitations of study* – this section is a discussion of those factors that constrained or limited your research. For instance, these may involve aspects of data collection, sample size, skews or biases in the dataset, and the relative quality of the responses (Chapter 12). It is an opportunity for you to show the examiner that you understand how parts of your research were affected. There is a common misperception among students that only dissertations that appear to have 'worked perfectly' will get the highest marks, and that those that encounter or report problems will be penalized by lower marks. This is not the case: one significant indicator of your ability to undertake independent research successfully is your capacity to manage issues that impact on your research.
- *Implications for future research* – this final part of the chapter is used by you to suggest several (i.e. two or three) potential research areas that grow from your dissertation study. Each requires a paragraph of discussion, not just a single sentence invoking them briefly, but giving the reader the basic substance of your idea. Further research is often required to overcome many of the limitations previous research (including yours) has encountered. Therefore, this and the previous section are related.

Many students struggle with this final task. These are not necessarily policy recommendations or practical implications (unless your programme requires them) which are more usual in consultancy and contract research. Rather, in the course of conducting your work – perhaps the design of the data collection, the analysis of the data, or the final writing up – new questions or avenues for research may be revealed. You should think of limitations and future research as connected in the same way as opportunities and threats in SWOT analysis: out of every threat you should attempt to make an opportunity. Here it is the same: limitations suggest potential improvements you could make in future research. For example, limitations, such as the sample size, the sampling frame/s used, or the case studies selected, may suggest that further enquiry is necessary to supplement the current work. Who knows, this may be research you want to conduct as part of your master's, doctoral or post-doctoral career.

There are other issues you may consider when reflecting on limitations or future research, as Oliver (2008: 137) notes. These include:

- Were there ethical issues that should have been addressed more thoroughly?
- Were there practical alternatives to the methods you ultimately decided to use?
- How might these have resulted in different types of data, results or findings?
- Is there any way in which the collection of data could have been enhanced?
- Could the validity and reliability of my current data have been improved?
- Are there any new studies that have appeared since I designed my work which may have inspired a different direction or approach if I was aware of them earlier?

Putting together your chapters

One of the most basic but hardest things to decide is which chapter should be written first and in what sequence should they be written. The obvious solution would seem to be to start with Chapter 1 and continue in a logical order through the chapters. As we noted in Chapter 2, this is routinely not the case and other common starting points are adopted. We do not wish to rehearse those arguments again except to reiterate that you should leave the conclusion and, in particular, the introduction until near the end. The reason for this is that under certain circumstances it may be necessary to adjust your research objectives, if for example, your data collection did not go according to plan. Under such circumstances your data may not be that useful for researching your original objectives. Assuming you have no time or resources to start collecting fresh data then the logical thing is to readjust you research aims, a point returned to in Chapter 14.

Keeping to your word limit

All dissertations have word limits which vary across institutions and even between different programmes within the same institutions. Typical word limits range from 8,000 to 15,000 words, and so a key skill is not to waste words and to know which parts of your dissertation to give more words to.

One of the most frequently asked questions by students is how long should particular chapters be? Whilst there is no hard and fast rule we can offer some guidelines from which to work. These are based on what may be regarded as your key chapters. As we explained earlier, all your chapters are important, but our advice tries to give you a more detailed focus in terms of word allocation.

The example we take is based on a 15,000 word dissertation, as the word counts for each chapter reflect the structure and content of chapters as discussed previously. In this context those chapters that have a standard structure (i.e. the introduction and conclusion) can be completed in a relatively small number of words, while other chapters such as the literature review, methods and results would require relatively greater numbers of words. Within this student's marking scheme, the literature review and the methods were heavily valorized, hence the greater space afforded to them. The student has also used two methods and hence requires 2,000–3,000 words to write that chapter. Who knows, there may have been an opportunity cost incurred here: if only one method had been employed, there may have been more scope for a longer literature review and/or data analysis? An alternative interpretation is that there may have been greater scope for a more intensive discussion of a single method.

Box 13.4 shows an example of this idea, and it assumes that appendices and lists of tables etc. are not counted in the total word limit. One piece of advice, which the student has followed here: in planning your writing, plan the length of your chapters in ranges (i.e. the introduction will be between so many words, for instance 1,100 and 1,500 words in this case). Here the student uses the word limit to shape the upper limits and their project word count, while the lower estimates total 80 percent of the available word count (i.e. 12,000 words). Students are often worried about exceeding their word counts. This is not

Box 13.4 A detailed structure plan showing the use of words across the chapters

Chapter	Heading	Page	Approx. word count
	ABSTRACT	i	Do not contribute to word count
	CONTENTS	ii	
	LIST OF TABLES	iii	
	LIST OF FIGURES	iv	
	LIST OF APPENDICES	v	
	ACKNOWLEDGEMENTS		
1.0	INTRODUCTION		1,100–1,500 words
1.1	Background	1	
1.2	Aims of the study	2	
1.3	Objectives of the study	3	
1.4	The case-study area	4	
1.5	Structure of the study	9	
2.0	LITERATURE REVIEW		3,400–4,000
2.1	Tourism and the tourist gaze	10	
2.2	The character of consumption	12	
2.3	Post-modern culture and consumption	12	
2.4	The consumption and growth of film	13	
2.5	The evolution of film-induced tourism	15	
2.6	The impacts of film on tourism	18	
2.7	The limitations of film-induced tourism	20	
3.0	METHODS		2,000–3,000
3.1	Research approach	22	
3.2	Case-study selection	22	
3.2.1	The Headland Hotel	22	
3.2.2	Prideaux Place	22	
3.3	Quantitative data collection – questionnaires	23	
3.3.1	Advantages and limitations	23	
3.3.2	Design	25	
3.3.3	Sampling	26	
3.4	Qualitative data collection – interviews	27	
3.4.1	Advantages and limitations	27	
3.4.2	Content	28	
3.5	Pilot study	29	
3.6	Methods of analysis	30	
4.0	DATA ANALYSIS		3,500–4,000
4.1	Introduction to data analysis	31	
4.2	Theme 1: patterns and demographic profiles	32	
4.3	Theme 2: the influence of film on tourism to Cornwall	39	
4.4	Theme 3: film as a motivator for travel to Cornwall	46	
4.5	Theme 4: the effect of film on tourism in Cornwall	53	

(continued)

Chapter	Heading	Page	Approx. word count
5.0	CONCLUSIONS		
5.1	Introduction to discussion	57	
5.2	Theme 1: patterns and demographic profiles	57	
5.3	Theme 2: the influence of film in Cornwall	60	2,000–
5.4	Theme 3: film as a motivator for travel to Cornwall	62	2,500
5.5	Theme 4: the effect of film on tourism in Cornwall	64	
5.6	Limitations of study	66	
5.7	Future research	67	
	REFERENCES	69	Do not contribute towards word count
	APPENDICES Appendix 1: movie maps of South West film locations Appendix 2: example of the questionnaire Appendix 3: Structure of interviews	76	Do not contribute towards word count

Source: adapted from Wild (2005)

least because their marks can be penalized (in some cases on a sliding scale depending on how far it is exceeded). As a rule of thumb, aim for the lower limit and you'll find that you have the flexibility and insurance if/when you need to write more. You should still finish within the word count: remember, this is the upper limit you should not exceed so it is perfectly acceptable to submit *below* the limit. Finally, in this regard, you should note that there is not a simple linear correlation between the length of your dissertation and its final mark. Some of the most compelling, elegant, well-argued, professionally produced dissertations we have read have also been some of the shortest. Remember, less can mean more and more can be a bore!

Constructing an internal structure

Your dissertation should be more than a collection of well-written chapters. It also needs to have what we will call a strong internal logic based around your research objectives. There are a number of ways of looking at this, including the need to keep a clear focus on your objectives at all times. For example, as we have explained previously, in the literature-review chapter the purpose is to examine the literature in terms of themes relating to your objectives. It is not just a general review of the literature. Similarly, in your methods the focus should be on the information required to research your objectives.

Within Box 13.4 you can see how the student has attempt to weave these strands of continuity through the text. The aims and objectives are discussed in Sections 1.2 and 1.3. From chapters 4 and 5 it is clear that the student had four objectives in this study (based on identifying who are film tourists, how important the influence of film was in making their journey, specific film-related motivations, and assessing the effect of film tourism more widely). Irrespective of the merits of the objectives, it is plain to see how they have shaped the data analysis (fourth chapter) and how the main findings feature in the conclusion (fifth chapter). Of course, this is a very explicit and mechanical approach, and it works for the majority of undergraduates we supervise. If you are a more confident writer, you may be able to weave those strands in a more subtle or creative way.

The chapter at a glance

The main learning points of this chapter are that:

- Your dissertation is comprised of a series of discrete elements, each of which fulfils a particular role in your final text.
- It is important to establish a strong internal logic to your dissertation by establishing connections running throughout the text by means of the aim/s and objectives.
- You must write with a purpose.
- You don't have to write the main text of your dissertation in a simple linear sequence from the first word of the introduction to the last word of the conclusion – find a sequence that suits you.

Dissertation checklist

Before you go further in your work, check you:

1.	Have included all the components normally expected of a dissertation.	
2.	Have a clear structure for your dissertation and for each chapter.	
3.	Have invested enough time in the introduction and conclusion as the last parts you are likely to write.	
4.	Are within the word limits prescribed by your institution.	
5.	Recognize the importance of signposting the reader through your dissertation as well as through each chapter.	

Part IV

POST-PRODUCTION

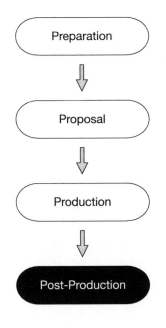

14

BEFORE YOU SUBMIT: SOME FINAL CONSIDERATIONS

Learning outcomes

By the end of this chapter you will be able to:

- Appreciate that style and format counts.
- Review your own work more effectively.
- Understand the value of back-ups.

The importance of 'post-production'

You are reaching the proverbial 'end of the line'. With your analysis conducted and well on the way towards completing your writing up, you may wonder what comes next? As soon as you finish the text, you may be tempted to rush to print your dissertation, submit it for assessment (Chapter 15) and move on to other things. Many students we have supervised felt that, having undertaken their analysis and write-up, they have completed their work. Sadly, this is not the case and, if you are unduly hasty and do not properly finish your work, you may run the risk of eroding your final mark. Remember, it is the little things that can make a big difference.

As the closing part of Chapter 13 indicated, better dissertations are those in which the sum is greater than the individual parts; that is to say, time invested in improving the internal consistency of your dissertation text is time well spent because it should result in a dissertation that offers a (greater) impression of a coherent, logical, well-executed piece of independent research. 'Post-production' – to borrow a term from the film industry – can take two (overlapping) forms, which are discussed here: effectively reviewing your work, and copyediting the final text to ensure accuracy and completeness. Both are somewhat time consuming if they are done properly and they may not be the most exciting tasks, but they are important, as you will learn below.

There is no doubt that this chapter is usefully read towards the end of your dissertation work as you complete the assembly of your final text for assessment. It will provide several reminders of tasks for you to undertake. However, you will find it more efficient to conduct many of the tasks described in this chapter, such as compiling the reference list or properly appointing tables and figures, along the way, perhaps in parallel to writing the main narrative of your dissertation. It is also unwise to leave everything to the end because you'll be surprised how rapidly the finishing line comes into view when you enter the home straight. Whereas Chapter 13 considered what needs to be brought together to form the final document, this chapter focus on refining it so it reaches the highest standard possible. 'Rounding out' your dissertation is challenging work; it's your last chance to maximize the impression you leave so it is critical to get it right.

Have you addressed your aim/s and objectives?

At the end of the previous chapter, we encountered the need for you to think critically about your dissertation with a view to establishing the limitations of your work as well as the implications of your research for future academic enquiry in your topic area. However, in that discussion we deliberately overlooked one of the most important questions in any auto-review, and one which may require you to undertake some notable revision of your draft dissertation.

That difficult, if not daunting question is, to what extent have you achieved what you stated you were going to do in the introduction of your dissertation by the time you reach your conclusion? Put another way, one of the first tasks in reviewing your work is for you to assess whether you have adequately addressed the aim/s and objectives you set yourself.

To do this:

*   Think carefully about your data. Is there enough *evidence* in your *data* to address your objectives or to support answers to your research questions?
*   Are there specific elements or particular features about your data that speak more (or less) conclusively to particular objectives and/or research questions as they were originally drafted?
*   Did your choice of *methods* and *analytical techniques* enable you to address your objectives or research questions adequately? At the end of their research some students find that they have not measured what they intended to, thus restricting their ability to deliver on their original stated goals.

You may decide that you have not addressed particular objectives or individual research questions to what you perceive as a satisfactory standard. Through an understandable fear of failure, some students panic at this prospect, thinking that they may have designed their research poorly or not asked the right questions in their survey instruments. As Oliver (2008) indicates, at this point it would be entirely appropriate to record your judgements as limitations of the research in the conclusion.

Nevertheless, in our experience students are reluctant to do this. Moreover, implicit in such a course of action is that your objectives or research questions would be addressed as they were originally stated. We have encountered many situations where students feel, for one reason or another, that this was not ultimately possible and in Chapter 13 we raised the prospect that it may be sensible to adjust your aim/s and/or objectives as a response.

There may be other reasons why this has not been possible for purely legitimate reasons. Possible areas to discuss with your advisor include:

- *Exploratory research* – if your research is exploratory in nature (as opposed to experimental – see Chapter 5), definitive answers are sometimes beyond reach. Your explorations into the nature of a particular problem or question may lead you to conclude both that additional research is necessary and/or the original objective may need to be redrafted.
- *Not enough data* – if you conclude that you may not have collected nearly enough data to be able to definitively answer the question you have set out, this does not mean that your research question is flawed. What you may be able to conclude is that, as a result of the work you have done, you have a better idea of how much work is necessary to answer the question (i.e. in a future repeat of the research). We may be closer to answering it as a result of your dissertation, but it needs a bit more work (and that work may be well beyond your current scope).
- *Reject the null hypothesis?* – if you have undertaken a study that attempts to address a research hypothesis and find that you cannot conclusively address whether or not to accept or reject the null hypothesis, a natural conclusion may be the call for additional diagnostics regarding research design (in order to ensure that you had drawn the correct inference after all, and not committed a Type II error). This would certainly be the case for a master's or doctoral thesis, but the limited time available during an undergraduate dissertation may preclude this. In this case, a full and frank discussion in your concluding chapter may be necessary. We have seen students in the past undertake more experimental-type dissertation research, only to conclude that their experiment was faulty. They subsequently spent a significant portion of their final stages of writing arguing why a better experiment must be constructed and how this might be done (i.e. in the section of the conclusion on implications for future research).

Before you submit: read your own work

A second form of review which may be slightly less daunting than the first, is to review your own text. Most people (perhaps with the exception of radio announcers or television personalities) absolutely loathe the sound of their own voice on recordings, so much so that they do everything they can to avoid hearing it. Perhaps not surprisingly then, many people also have similar problems with reading what they have written. Doing so produces feelings of anxiety and apprehension, and it has not been uncommon for us to hear from former students that they had no interest in reading what they had written after they have completed their first full draft.

A fear of reading one's own written work produces several problems. Firstly, unless you are enlisting the services of another person to proofread, your work will go largely unchecked. This is an especially risky strategy, particularly if English is not your first language. Some examiners are quite picky when it comes to spelling, grammar and the general appointment of a dissertation. This is perhaps understandable in the sense that the dissertation is supposed to be the pinnacle of academic achievement – it is the last thing a student does before graduation – so the standard of (textual) appointment may be an assessment criterion (Chapter 15). Need we recommend you use the spell checking and grammar functions in your word-processor?

Secondly, an unwillingness to review your writing means your abilities and skills as a writer will take longer to improve. Thirdly, by not reviewing your work you deprive yourself of a chance to enhance what you have written. Revisions are a critical part of the writing process, and often new ideas and fresh perspectives develop from revising a text. Assuming you are committed to your text being reviewed, we can offer the following tips to help:

- *Review both piecemeal and in totality.* You should review your writing on a regular basis. This can be daily, weekly or at significant milestones (for example, finishing a section of a chapter). You should also review your entire dissertation when completed. Read it like you would a book, from front to back. Do not be afraid to be critical, and prepare yourself for lots of necessary changes once you have finished your review.
- *Avoid the temptation to review large chunks (i.e. a chapter or set of chapters) soon after you have written them.* Let your writing sit for a while, at least a week if possible, while you work on other aspects of your dissertation. This will help you cast a fresh (and critical) eye over your work, and you may be more inclined to identify problems as a result.
- *Be objective and balanced*, or at least as much as possible. This is perhaps easier said that done, but when some people read what they have written they have a tendency to be their own worse critics. On the other hand, we have seen some students who are far less critical of their own work compared to their views of others' writings.
- *Have someone else do it.* Having a fresh pair of eyes review your work can be hugely advantageous, and is especially useful for those where English is an uncomfortable second language. Your reviewer will undoubtedly find things that you have overlooked. At the risk of stating the obvious, be sure to utilize someone who is not related to you, is not in love with you or owes you money! Use someone at arms-length from your work, and be sure to reward them. It does not matter if they do not understand the subject of your dissertation – in fact, the less they understand the better in many cases – because their task is to assess readability, phrasing and logic.
- *Do not overdo it!* The 'law of diminishing returns' is perhaps usefully invoked here: there is no simple linear relationship between the amount of time you invest reviewing your text and your final mark. If you are not careful you may spend a considerable amount of time revising what you've written, only to realize that you have not changed the fundamental substance of the argument. Granted, the editing phase of one's work can be long and tedious, but it is easy to fall into a trap of near constant revision. You have to be pragmatic and know when to draw your revisions to a close.

Few documents are truly error-free, and all documents can be improved upon. Some revisions are necessary (grammar, format, etc.) while others, such as finding new ways to say something that is already stated clearly enough, are inefficient.

Visual cues and structure (or, style and appearance is everything)

There is nothing more frustrating for an examiner than having to review a dissertation that, although intellectually very stimulating, has been presented very poorly. Fortunately, fixing this is very easy. Researchers and writers talk of production values in the same way as a film producer: we all strive to make the final product visually appealing and to improve the overall look and feel of our work.

Beyond the five chapters that provide the main substance of your dissertation (see Chapter 13), there are several other components you must complete correctly (see Box 13.1).

References and appendices

Make sure you have listed all the references mentioned in your text – the usual system is the Harvard system as explained in Chapter 4. It is surprising how many dissertations have missing references (which can be a reason for losing marks). This 'list of references' appears immediately after the conclusion. Some students (erroneously) label this as their 'bibliography'. In fact, a bibliography is a list of all materials potentially pertaining to a topic which you may (or may not) have consulted or employed in your study. It is very unlikely that you will be asked for this. Rather, you will be required to provide a listing of all of the works cited in your text (i.e. a reference list).

The process by which you build a reference list for a document like a dissertation can be onerous and at times frustrating. If you like to build your reference lists manually, it is best to have a separate 'references' file open at all times while you are writing. That way, when you make reference to a source in the text of a specific chapter, you can immediately switch documents and add the reference. Forgetting to do this represents a significant downside. Another is that you must spend quite a bit of time when you complete your dissertation manually checking to ensure that all references have been included. This can be time consuming and is prone to mistakes, and it is one reason why this task should not be left to the last minute.

Technology can rescue you from this petty torment. Many popular reference management applications such as EndNote (www.endnote.com) for Windows and Sente (www.thirdstreetsoftware.com) for the Mac automate reference generation. These applications work seamlessly with most word-processing applications (including Word) and are generally easy to use. Check with your institution to see if site licences are available for these and other applications as they may save you considerable time.

Appendices are an extremely useful means of putting information in your dissertation that is useful background which informs your discussion of key points but which would overload the main text by exceeding the word count and/or disrupting its flow. This information is supplied separately at the end of the dissertation after the reference list. For instance, copies of your questionnaire and interview schedule (as well as covering letters and consent letters) should be placed in appendices along with an example of a completed interview (i.e. anonymized transcript) to illustrate the full quality of this type of information. All these additions demonstrate to your examiners that you know how to prioritize, present and organize your material, as well as the fullness and correctness of the procedures you have undertaken.

Like the reference list, appendices do not normally contribute to the word limit for your dissertation. Finally, make sure that your appendices are numbered sequentially (i.e. Appendix 1, Appendix 2, etc.) and that they have appropriate titles. You should also compile a 'list of appendices' to appear in the 'preliminaries' (see below). Moreover, the reader should be referred to each appendix at some point within the main text. Some students make the mistake of thinking that readers will somehow make the connections for themselves as to when and how an appendix should be consulted while they are reading the main narrative.

Appointment of the text

By the time of its submission, you should aspire to the most professional-looking dissertation possible. As a first step, ensure that the margins are even throughout. We have seen countless examples of students who write on different word processors throughout their project, only to find that, when they merge documents, the margins are different from document to document. This produces uneven text rendering and can be visually distracting. We recommend using one word processor only for your work and ensuring that all margins are equal.

Matters of text justification and spacing often arise when the final dissertation is about to be printed. Some institutions stipulate full justification (i.e. left and right margins) while others prefer only the left margin only be justified. Similarly, some institutions demand double-spacing of the text while others will accept 1.5 spacing (single-spacing is usually excluded).

For the definitive answer on how to produce your text, check your institution's regulations, which also provide you with guidance concerning the often vexing production issue of headings and sub-headings. First and foremost, they must be consistent. Whether you use numerals (14i, 14ii etc.) or decimals (14.1, 14.2 etc.) or text embellishments such as **boldface** or *italics* is less important than actually being consistent throughout. If you are free to structure as you see fit, you should consider adopting a policy where there is minimal use of headings and sub-headings. A document with more sub-headings than necessary (say beyond a second level of sub-headings i.e. 14.1.1) can appear disjointed and overly compartmentalized. One common criticism of such dissertations is a lack of flow.

Consistency of appearance is also important for figures and tables. By figures, we mean any graphic (i.e. vector or raster image) whether a diagram, map or (digital) photograph embedded in your text. With respect to tables, the careful placement of lines (a feature found in most word-processing applications) results in a very clean look with little confusion over column and row configuration. Box 12.1 demonstrates a simple but interesting table design and formatting (notwithstanding our comments previously about over-reporting), and this is consistent with the colour scheme in the chart (Box 12.1). Together they present an impression of competent and professional production of an unwaveringly high standard. Quite often, there is the temptation to include all lines around all boxes in a table, but you need to exercise some discretion. At times this can result in rather awkward and 'busy' tables.

As we have noted in Chapter 12, it is important to consider the presentation of your data carefully. Many software packages offer the ability to produce attractive charts, but it can be confusing as to what type of chart best matches the data you have generated. In a similar vein, you should use a consistent colour and/or shading scheme for the graphics throughout your dissertation. Often the colours or shades used by software will be based on a series of arbitrary defaults. The result can be a dissertation of many colours. For some this may be attractive; for others – probably the majority of readers – a preferable, not to say professional-looking, text will use a consistent (and relatively limited) colour-scheme throughout.

You should adhere to local instructions but also think about your own ideas on graphic design and visual identity within these parameters. Figure 14.1 is produced in the same format as Box 12.1 to emphasize the impact that can be made if you harmonize the production of graphics (and tables) throughout your dissertation. In this case, a greyscale colour scheme was used because the dissertation was going to be printed by a very high-quality laser printer. Old-fashioned stripes and hatchings can be helpful, crisp and clear if your printer has problems in accurately rendering the differences between shades (of grey). If you are going to use colour, make sure that your printer has adequate ink to

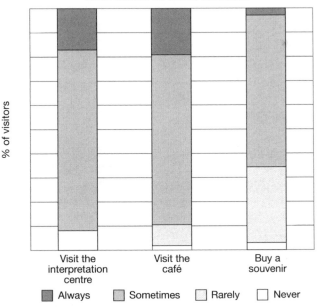

Figure 14.1 A high-impact graphic from a hypothetical dataset

Source: authors

print the colours faithfully and consistently. We see dissertations where the ink has been clearly running out!

Furthermore, you should ensure that all figures and tables are:

- Adequately described in their titles in terms of their content and purpose.
- Consistently and consequentially numbered in the text.
- Properly sourced and attributed.
- Mentioned and/or discussed in the text.
- Located as close as possible to their first mention within the text.

In the case of figures, you must ensure that all axes or categories are appropriately labelled. Do not accept labels that are automatically generated by your software package without reviewing them: they mean something to you because you will intimately know your data-set, but there is no guarantee that they will be obvious to your examiner. In Figure 14.1, the hypothetical title (as it might appear in the dissertation) has been included on the chart area (because our proper title is somewhat different). By and large we would advise that you don't do this because you have much greater flexibility in revising and renumbering titles if they are not part of a graphic that has been pasted into your manuscript.

Any figures or tables that use direct reproductions should have their source/s attributed and you should be careful not to convene any copyright arrangement or licence agreements. For instance, the maps and datasets that underpin the use of geographical information systems (GIS) are often made available to users in higher education on preferential terms (from other purely commercial users) and their use has to be appropriately acknowledged.

Maps for dissertations are routinely sourced from the Internet these days, and simply pasted into the text without attribution of source (which is incorrect, if we need to remind you). Make sure the quality of the map is acceptable. Nothing appears less professional than a map with a very poor resolution, resulting in a blurry appearance on the printed page. You should also note that, technically, a map is not really fit for purpose if it lacks a scale and north point. For the purpose of high-quality printing, your images – whether from digital photos, scans, screenshots or downloads – should be saved at 300 dpi (dots per inch) at least. Images can be resaved at a higher resolution in most photo-editing software.

If you have adapted a diagram from the original or modified a table (perhaps by removing columns and/or rows, or by calculating additional figures), you should acknowledge this by reporting 'Source: adapted from ... (year)' or 'Source: abridged from ... (year)'. If the table or figure is a result of your thinking or synthesis of ideas, it may be attributed as 'Source: author' or 'Source: author adapted from ... (name of original source/s)'. Your data in these forms would be attributed as 'Source: author's fieldwork' or 'Source: author'. The key point is that you leave the reader under no doubt that you are properly attributing the origins of the material you use and that you are not passing off another's work as your own. Should we need to remind you, in so doing you will be working to avoid any possible accusations of plagiarism.

For mathematical formulae, if you are using Microsoft Word be sure to use the 'equation editor' that is normally built-in to local installations of the application. If this is not available, check with your local information technology service on campus to see what options are available.

Preliminaries

It might seem somewhat odd that the first elements a reader encounters in your dissertation (see Box 13.1) are the last things you have to complete before your dissertation is ready for submission:

- *Title page* – this should include the full title for your dissertation and it will usually require you to enter a standard form of words declaring that you alone are the author of the work, your dissertation is an original piece of work, and that you acknowledge that you have read and understood your institution's rules concerning plagiarism.
- *Abstract* – many programmes require you to enter an abstract. Very similar to an executive summary, this is supposed to be a very basic synopsis of the work and will probably not extend more than one side of A4, detailing not much more than the aim/s of the study, the basics of your methods and analysis, and the most important findings and conclusions.
- *Table of contents* – this lists all chapter and section headings and sub-headings, their titles and the page numbers on which they start. It includes preliminaries as well as the addenda to the dissertation. The preliminary pages are normally numbered by Roman numerals while the main text, starting from page 1 of chapter 1 use Arabic numerals.
- *List of tables*, *list of figures* and *list of appendices* – all tables and figures have to be numbered. Some institutions prefer them to be listed from 1 to n throughout the main body of the text, while others stipulate that they must numbered consecutively by chapter (e.g. the two diagrams in chapter 3 – 3.1, 3.2 – are followed in the next chapter by 4.1, 4.2, 4.3 etc.). Check which system your institution uses, number accordingly and then ensure that all tables and figures, their numbers and page locations, are entered into lists that follow the table of contents. The number, title and page reference for the start of each appendix is provided in the list of appendices.
- *Acknowledgements* – not a requirement, but if you want to thank all and sundry who've been helpful along the way, this text usually goes between the list of appendices and the start of chapter 1: introduction (after a list of abbreviations if, likewise, you have any need for one).

Preliminaries are page-numbered, but instead of using Arabic numerals (1, 2, 3 etc.) as you would for the main body of the text, including the references and appendices, you should use Roman numerals for the preliminaries (i, ii, iii etc.), usually starting at page i for the abstract (i.e. the title page, frontispiece, has no number whatsoever, and your introduction starts on page 1).

And finally with respect to the preliminaries, it is perhaps worth reiterating that the statement on the title page is essentially a form of 'plagiarism declaration'. In using this form of words and signing it, you are reassuring your institution that the work is yours and that you have employed the highest level of academic integrity in its collection.

Page setting

Most dissertations are written in Microsoft Word, Apple's Pages or some other similar word-processing system. If you consider yourself technologically savvy, or are interested in learning a bit more about document processing, you should consider using a professional typesetting system such as LaTeX.

LaTeX is freely available (www.latex-project.org and www.tug.org) and works on all computer platforms. The benefit of LaTeX is that it produces beautifully typeset documents. It allows you to focus more on writing rather than formatting. A downside of LaTeX is that there is a reasonably steep learning curve, so it is perhaps reserved for those who plan to continue in education and/or will use it after graduation.

Box 14.1 And then there was none? Dealing with disappearing data

Using the Internet for your research can be a double-edged sword. The web provides you with opportunities to reach many people and to receive their responses quickly in a format that can be efficiently processed and uploaded into analytical software. It can also be a cruel and unforgiving medium. At a single key stroke data can be deleted, never to be retrieved if you haven't put in place adequate measures to safeguard your sources.

One of our students encountered a novel variation on the 'all or nothing' theme. In addition to a questionnaire survey which had been conducted online, a Delphi consultation of tourism experts and their opinions had been conducted by e-mail with the results stored on the student's server space. Now we can see what you might be thinking and you're right: the student should have backed up the material but not everyone is IT savvy.

There is no denying that a harsh lesson was learned. However, all was not quite as you might think. The entire dataset didn't disappear, just half of it through a critical server error. About 45 percent of the questionnaires remained and about 15 percent of the final Delphi responses were lost. The crucial question faced by the student was how to proceed? Continue with the remaining data? Start again? Perhaps even run a further phase of data collection online to augment the remaining questionnaires? It was not as though the whole lot had disappeared and it was necessary to start from scratch.

The student discussed several options with advisors. Time was against starting again and anyway, emotionally the student didn't want to write off the remaining data which had been a struggle to collect. Additional data didn't seem to make sense and how could a robust intellectual case be made for this? So, it was decided to soldier on with the remaining data. The objectives were adjusted to reflect the new balance of the dissertation. After all, the majority of the Delphi data had remained, as well as the student's notes and preliminary analysis. The remaining questionnaire results were analysed. For the write-up, the limitations were obvious as were the implications for future research. But what would *you* have done in these circumstances? Let's hope you never have to face such a dilemma.

Source: authors

Avoiding apocalypse now: technology for backing up

We presume that, by now, you will have established a process by which you are regularly backing up your work to avoid the obvious risks of data loss (see Box 14.1). If not, put this book down and immediately make three digital copies (and this does not include the original that exists, most likely, on your computer) of the folder on your computer that contains your dissertation and all associated files. To maximize your back-up strategy and minimize potential loss, you should utilize separate media for your back ups. One back up, for example, can be on a CD or DVD. Another can utilize an external hard drive. External storage on portable hard drives or small USB 'pen drives' is generally cheap these days, with 1TB bare hard drives usually reasonably priced at many online retailers.

Now that you have a full back up of your dissertation, you are not quite finished. Take one of those digital copies to a relative or close friend's house. This becomes your off-site back up. While you may feel this is overkill, off-site back ups are invaluable in case there is a significant problem (fire, robbery, earthquake) that wipes out your master copy and all other back ups. After you have made your three copies and have taken one to an off-site location, you should consider putting yet *another* copy online, as cloud storage is becoming ever cheaper. Amazon has a service (S3 or AWS) that makes use of its vast array of servers around the world. You are charged for the data upload monthly. Other services include DropBox (www.dropbox.com) and SugarSync (www.sugarsync.com), both of which offer small amounts of free storage but which could be more than enough for your work.

Getting geeky about back ups: a quick note on version control

When writing up your dissertation, you will undoubtedly go through many versions of a document. This is to be expected and all writers face the prospect (sometimes daunting) of managing numerous versions of the same document. Thankfully, help is available, but it does require a bit of effort. Strictly speaking, version control (VC) when implemented on a computer used for writing, represents a process by which you can 'roll back' to a previous version of a writing project or file. For example, let us assume that you have separate files for each of your chapters in your dissertation. They may look like this:

Chapter 1 – Introduction.doc
Chapter 2 – Literature review.doc
Chapter 3 – Methods.doc

. . . and so on.

What many students do (although quite a few still do not) when they have finished writing a significant section of a particular chapter is to resave the file with an appended date and time to the file name itself. For example:

Chapter 1 – Introduction 2 June.doc

So what they are eventually left with is a long list of 'introduction' files:

Chapter 1 – Introduction 2 June.doc
Chapter 1 – Introduction 5 June.doc
Chapter 1 – Introduction 7 June with different sig section.doc
Chapter 1 – Introduction 9 June removed section 2.doc
Chapter 1 – Introduction 10 June sent to advisor morning.doc
Chapter 1 – Introduction 11 June.doc

Generally speaking, having multiple versions of any single file is not necessarily a bad idea. It allows you to move back in time and see what major changes you have made. The problem with this is you end up with a long, messy list of files and, potentially, a confusing (or worse, incomplete) record of what was changed. While you can certainly indicate some information on what was changed in the actual name of the file itself (e.g., 'Chapter 1 without section on rationale.doc'), sometimes you may find yourself wanting to be more specific about the changes you have made. Moreover, long files names with lots of text are cumbersome and not terribly efficient.

One way to alleviate the confusion associated with the differences between versions of a particular file is to employ a version control system (VC for short, and sometimes called 'revision control'). VC systems come from software engineering and are particularly useful when more than one individual is working on a particular project. Over time, researchers and other writers came to realize that a VC system can be just as beneficial for their individual work. The exact mechanics behind a VC system are beyond the scope of this book, but thankfully there are many online tutorials that explain the process step-by-step.

In a nutshell, a VC system means that, anytime you undertake changes to a particular file (for example, you have added a significant paragraph, or deleted an entire section), you tell the VC system about that change (generally called a 'commit') and, at the same time, provide details about what changes you have made. The name of your file never changes, and you never have more than one file, yet the VC system keeps an accurate record of the changes in the file. Most importantly, it allows you to revert back to any previous version of your file.

Say, for example, you were working on your methods chapter and, for the past few weeks, you inserted and worked on a specific section that outlines methodological approaches to tourism research. After those few weeks, you may decide that this new section does not work and you want to revert back to a state where methodologies were discussed more generally. With a VC system, this is easy: you simple tell it to revert the state of the file back to that point in time before you started the new section.

A word of warning, however: using a VC system is not always the most user-friendly experience you will have with your computer, and some say that it is reserved for those who are more technologically savvy than most. In actual fact, the absolute basics are usually very easy to learn, and once you master it you will revel in its ability to track the changes made to your document. Git (git-scm.com) is often cited as the most robust VC system, although others (such as Subversion, http://subversion.tigris.org) are also available. Best of all, to use the bare components of these systems does not cost any money. Git, for example, is entirely open source, and thus freely available.

A final word

In this chapter we have attempted to show you how to finalize and polish your dissertation to the point where it shines as a research document. Now it is ready to be submitted and you are finally able to reward yourself, for you have joined the ranks of researchers who have spent countless hours formulating aims and objectives, undertaking their own data collection and analysis, and assembled everything into a coherent and engaging document. For some of you, the dissertation will represent the very last thing you will do in higher education. For others, it is a mere stepping stone to additional study and research. Either way, you have joined the ranks of scholars who, over the centuries, have produced knowledge and sought clarity through research.

The chapter at a glance

The main learning points of this chapter are that:

- Appraising the extent to which the original intentions for your research are achieved, is an important first step in finalizing your dissertation.
- The time you invest, and attention to detail demonstrated, in post-production can have an important effect on how your dissertation will be evaluated.
- Reviewing your own work is an extremely useful exercise in the context of your dissertation as well as a wider skill to foster.
- Back-up your work on a regular basis to minimize lost time if you experience data loss.
- Version control can be a helpful device in shaping your dissertation by managing drafts.

Dissertation checklist

Before you go further in your work, check you have:

1.	Appraised whether you have adequately addressed your original objectives (and answered your research questions).	
2.	Reviewed each chapter individually and your dissertation as a whole.	
3.	Appointed your text (i.e. margins, justification, headings and references, tables, figures, appendices) consistently throughout.	
4.	Made sure that all of the references in your text appear in your reference list and, likewise, that all of the references in your reference list actually appear in the text.	
5.	Reread your institutional guidelines on plagiarism to eliminate any risk that you may contravene your institution's position.	
6.	Backed up your dissertation.	

15

JUST WHEN YOU THINK IT'S ALL OVER . . .

Learning outcomes

By the end of this chapter you will be able to:

- Identify the model of assessment used in your dissertation module or course.
- Understand how different dissertation assessment schemes function.
- Identify the different readers of your work and their roles in assessment.

Wait, there's more?

So, you've completed your dissertation, it's been bound, and you have handed it in. Your work's in the safe hands of administrators. Time for a huge sigh of relief? You've done your bit and now it's down to the examiners? This is one chapter of your life that has come to a close?

Well, perhaps not. You may have submitted your dissertation by the deadline, but there may be one or two final tasks still to complete! No doubt you'll be interested in what happens to your dissertation after you've submitted it. Most students we advise want to know how their dissertations will be assessed and when they will know the result. This is hardly surprisingly in light of the contribution of the dissertation to their overall result and the possible importance to life afterwards.

How will my dissertation be assessed?

Like any other modules or courses on your degree programme, the dissertation has a series of intended learning outcomes (ILOs). This is academic jargon for what you are expected to know or be able to do by the end of your dissertation studies (i.e. once you've handed it in). The ILOs should be stated in the documentation supporting your dissertation alongside the marking scheme or 'rubric' in North America: the latter is essentially a set

Table 15.1 A hypothetical marking guide for undergraduate dissertations

Class	Grade	%	Description
First	A	70–100	Excellent dissertation. Interesting aims, clearly set in previous literature. Evidence of original and independent thinking. Appropriate and accurate data analysis. Sound conclusions. Insightful discussion. High-quality reasoning, organization and presentation.
Upper second	B	60–69	Very good dissertation, meeting all requirements and most at a high standard. Falls short of first class on a few criteria or may be limited in some minor way.
Lower second	C	50–59	Flawed in one or more areas but exceeds basic requirements. As such, likely to be inconsistent. Level of detail, analysis or presentation may be uneven. Breadth of reading, conclusions or evidence of insight limited.
Third	D	40–49	Does little more than meet basic requirements. Research problem, literature review and methods may be weak or patchy. Poor presentation. Little if any evidence of originality or insight. Few if any hallmarks of good dissertation.
Fail	E	0–39	Broad category for several types of dissertation, including those not meeting institutional requirements (i.e. plagiarism).

Source: adapted from Parsons and Knight (2005: 147)

of grades and thresholds designed to guide your examiners on how to describe your level of achievement with respect to the ILOs.

An example of a possible marking guide for a dissertation is provided in Table 15.1. This is for a model of assessment based on the text alone (see below on p. 216). Adjectives and descriptions will naturally vary from institution-to-institution, but there are two important points to note here. First is the use of antiquated language in UK degree awards. Unlike other international systems of higher education, terms are used to categorize a range of grades. This is just a more traditional and colourful way of adding a label to grades than abstract letters or numbers in the case of 'grade point averages'. Still, for those not used to the UK system or for those studying in English as a foreign language, these may seem a little baffling at first. In the case of the latter, no doubt that by the time you complete your dissertation you will have learned what these terms mean and how they are defined in practice. If you are still unclear, ask your advisor/s to explain them in the context of your dissertation assessment. Indeed, this is an important general point for all users of this book, namely: you should ensure that you calibrate your understanding of grading systems and how they function. Your institution may keep a library of past work and your advisor/s may be able to point you towards the best or better examples, without of course divulging the marks they received.

You may have spotted the second important feature already based on the advice we gave you back in Chapter 1. There we set out to debunk the 'all or nothing myth'. As you will see, only excellent (i.e. first class) dissertations have few (minor) limitations, if any. Put another way, it is possible to achieve a very good (i.e. upper second class) or good (i.e. lower second class) outcome where some parts of your dissertation work perform notably stronger than others.

It is useful to reiterate this point. Many students falsely believe that their dissertation is only as strong as its weakest part; that is, if one component is weak then so too, by implication, is the entire dissertation. This is clearly not the case. It is pretty easy to dwell on one's weaknesses and limitations, rather than to focus on one's strengths and abilities. Such a negative mindset is not typical of how dissertations are assessed nor how professional research – either within or outside higher education – operates these days. Teams of researchers with complementary skills are assembled to overcome skill and knowledge gaps among individual employees. Effective researchers are those who have equipped themselves with the right skills, knowledge and experience in order to be able to conduct their research to a high standard, but who are also able to recognize the limits to, and gaps in, their competencies and to seek appropriate future training opportunities to fill them.

The vast majority of, if not all, programmes recognize this ethos. They adopt multiple assessment criteria, and they make their pronouncements based on the balance of the evidence. Table 15.2 demonstrates how this works practically with an articulated dissertation marking scheme based on five clear marking criteria. This scheme is interesting because it attempts to provide more differentiation than most. For instance, notice how it attempts to differentiate levels of excellence in the 'first class' range based on sustained high-level achievement. Also, there is greater differentiation in the range of marks from 50 to 69 percent compared to Table 15.1. It is unusual to find a marking range that bridges degree classes but this denotes the difficulty in categorizing dissertations at the class intervals and avoids a 'cliff edge' effect.

Clearly, if your dissertation is unwaveringly excellent in all aspects, you should receive the highest grading. Well done! Great Job! Your success is richly and rightly

Table 15.2 A possible marking scheme for tourism dissertations based on five criteria

Percentage	Class	Topic	Methods	Analysis	Discussion	Presentation
80+	First	Excellent in all criteria				
70–79	First	Excellent in three criteria, at least very good in others				
64–69	2:1	Very good in at least three criteria, good in others				
57–63	2:1/2:2	Good in at least three criteria, moderate in others				
50–56	2:2	Moderate in at least three criteria				
40–49	Third	Weak in at least three criteria				
0–39	Fail	Unacceptable in three or more criteria				

Note: 2:1 = upper second class, 2:2 = lower second class (see Table 15.1)

Source: authors

deserved. Conversely, if your dissertation is flawed in every regard, it is unlikely to pass muster. Not only are you unlikely to have read this book properly, but also you have probably not listened to your advisor/s or read your institution's guidance.

Many dissertations fall somewhere between these two extremes of performance. Some parts of your dissertation work will have been more successful than others; some sections or individual chapters will work better than others. So, the more consistently you have performed at a higher level, the greater your chances of receiving a higher final grade.

Different models of assessment

In much of this text we have implied that dissertation modules or courses are examined purely on the basis of the text you submit. In many higher education institutions around the world this may indeed be the case. However, there are different models of assessment for dissertation modules and to some extent these reflect the way in which independent research is viewed by each institution.

- *Dissertation only*

Perhaps the most traditional form of assessment, the text you produce is alone the subject of the final (summative) assessment and grading for the module or course. All the preparatory and background work you have done (i.e. in terms of proposal, ethical clearance, risk assessment) does not contribute separately to your grade; that is, they are formative assessments along the way that only contribute in a more indirect sense to the success of your final text.

- *Dissertation and research proposal*

In this model, research is viewed more as a process and your competence as an independent researcher is better assessed not only in the final outcome but also in your ability to conduct tasks along the way. For instance, your institution may believe that part of your assessment should test your ability to generate a 'researchable' topic (Chapters 3–5) and to devise a feasible programme of research (Chapters 6–9). Hence, here both your preparatory work and submitted text contribute to your final grade, although the former is likely to be weighted less than the latter. An example of a hypothetical articulated marking scheme that could be used in this regard is presented in Box 6.6.

- *Dissertation and presentation*

You may be asked to present your dissertation to an audience of fellow students and/or staff members (i.e. after the text has been submitted). Your presentation and text contribute to your final grade, although the former is likely to be weighted less than the latter. The thinking behind this model is that it is vital for effective researchers to be able to disseminate their work verbally, as well as textually. Your presentation may take place either individually, in front of panel of faculty, or perhaps as part of a symposium or showcase for your peer group to share its work. In either case, you should be sure of the focus: should you talk about your experience of the process and/or the major results and findings?

- *Dissertation, proposal and presentation*

A logical combination of the two previous permutations, especially if the presentation is at the end of the design phase (i.e. once your research proposal is complete). In this case your presentation is an ideal opportunity to gain value feedback and comments from your audience on your intended work in its current form.

There are relative merits in each model and it is important that you understand from an early point what is being summatively assessed, when and its contribution to (i.e. weighting in) the final grade, especially in the case of the latter three models.

With regard to the first model, there is no room for complacency. Some students view it as a charter to be cavalier or to make mistakes early on because, they think, it will not impact on their final mark. As it is non-credit-bearing, it is not important to invest the proper time in preparatory work. In our view, this is worrying. As we discussed in Chapter 6, this is a false economy, the consequences of which only manifest themselves several weeks or months down the line when the dissertation text is assessed.

More worrying for many students, though, is that they may be required to present their proposals to a group of fellow students or to conduct a short face-to-face interview with their advisor/s and/or a panel of other staff (i.e. faculty) members. This latter situation is more commonly the case at master's level, especially in North America. This anxiety often relates to the weighting afforded to the exercise. The proposal, the presentation and/or the interview may contribute as much as between 10 percent and 25 percent of the final mark for the dissertation programme. The logic behind this sort of approach tends to be that effective design of research projects is a key skill, as is the ability to articulate these ideas.

We would broadly endorse this view, but it is not always necessary to grade every single aspect of the dissertation process, for quite legitimate reasons. For instance, the proposal may function as an uncredited (formative) learning experience in recognition of the iterative nature of the research process. The proposal is, in effect, your chance to set out and discuss your work 'in progress'. The nature of the submitted dissertation is sometimes quite different to how it was originally intended at the time of the proposal, and students should not be prematurely judged before, or even penalized for, the natural evolution of their work. Professional research in the social sciences does not always go entirely as planned and without issues or complications (Law 2004). Rather, it is routinely punctuated by events, episodes and contingencies that cannot always be fully anticipated. Hence, your ability to react and adapt is critical to the success of your dissertation and these are both key attributes of the most effective independent researchers.

What makes a good dissertation text?

This is the golden question and there is no single right answer to it. One simple answer is, a dissertation that addresses the assessment criteria of your programme at an unwaveringly high standard. Thus, it is clearly in your interests to obtain and study the assessment criteria at your institution as well as a blank marking sheet, if you are not issued them before or during your studies.

The best dissertations – irrespective of subject or discipline – tend to be characterized by the following hallmarks (Parsons and Knight 2005: 145):

• A high standard of presentation, and easy to read.
• A clear problem set in the proper context (i.e. from the academic and/or grey literature/s).
• A clearly explained and appropriate method.
• Appropriate, systematic and comprehensive data and analysis.
• Clear separation of results and interpretation.
• Sensible and penetrating discussion.
• Logical and relevant conclusions.
• Intellectual achievement and originality.

In general, poor dissertations tend to have the polar opposite characteristics of the best dissertations. Absent from this list is separate mention of evidence of critical writing or critical enquiry. Often one hears that the best dissertations are the most critical. Within your dissertation experience your advisor/s may have encouraged you to be critical in your decision making and in your writing just as we have here. You may even see this as a separate assessment criterion. However, we would simply note that, in order to have developed these characteristics to a high standard, you will have had to have made clear, reasoned decisions on the balance of the information available for each task. These are the hallmarks of critical research.

With respect this listing, we quite deliberately placed production as the first point. To be clear, we don't believe in image over substance, but first impressions do count! Your examiners will have several, if not many other, dissertations to assess. Done properly, this is an arduous task that takes considerable time. A well-presented, professionally produced document cannot be bad if it puts them into a positive frame of mind before they've even read a word that counts. As Parsons and Knight (2005) correctly point out, at the first glance examiners typically inspect the title page, read the abstract, look through the contents page, flick through the main body of the text, and look at the reference list. They may even start reading the conclusion if it is brief. Both the reference list and the conclusion give the reader a first significant impression of the study and the direction it is likely to take.

As is the case with the assessment of research proposals (Box 6.6), it is highly possible that articulated marking schemes of some sort may be used by the examiners to assess your document. The sorts of headings that might appear on a marking scheme (like Table 15.2) or a marking form (like Box 15.1) are presented in Table 15.3. Some of these schemes are quite strict in scope and very mechanically generate a final mark based on the total marks on offer for each category. Other schemes are more flexible. In this case, as Box 15.1 reveals, an indicative grade is entered for each category and these are used to inform a single, final overall mark which is justified by the comments. For reasons of space, the comments in this hypothetical example are somewhat truncated, but they give you an impression of the sort of narrative that might be entered by one of your markers.

Here the student has been awarded a final mark of 64 percent for the document, or a mark in the upper second class in the UK scheme (Table 15.1). Overall, it seems that this

Box 15.1 A possible completed articulated marking scheme for a dissertation project based on tourism and disability

Overall mark 64%

Criterion	Indicative standard					Marker's comments
	70+%	60–69%	50–59%	40–49%	<40%	
Topic		✓				Well located in the (limited) literature. Disability is an overlooked theme academically despite policy-relevance. Not always willing to critique existing studies or policy. Clear aim and objectives.
Method		✓				Appropriate mixed methods strategy. Strong defence of questionnaire and clear rationale for combining with interview but explanation of interview programme could have been much better.
Analysis		✓				Competent use of descriptive and inferential stats. Sensible analysis of interview data. Not always integrated data in the manner expected of mixed methods research. Logical conclusions drawn from analysis conducted.
Discussion			✓			Didn't always link the results with the literature informing this study. Findings were somewhat limited therefore. Didn't address objective five adequately.
Presentation	✓					Excellent. Clearly written. Concise. Very well referenced. Diagrams to a very high standard.
Overall comment	A very good dissertation that demonstrates a high level of competence without being excellent. The topic clearly has merit and the research problem was well-positioned in a limited body of knowledge. The analysis didn't demonstrate the hallmarks expected of an excellent dissertation with one objective largely overlooked.					

Source: authors

Table 15.3 Possible marking criteria for dissertations

Criteria	Possible particular foci
Topic selection	Identification, definition (including aim/s, objectives, research questions), worthwhile rationale
Literature review	Quality, level of critical discussion
Methods	Appropriateness, implementation
Data	Collection, quality
Analysis	Appropriateness, accuracy, precision, depth of results
Discussion	Quality of findings, links to literature, logical conclusions
Presentation	Standard of text, appearance, use of terminology, writing style

Source: authors

student's work has been of a consistently high standard without being exceptional. Specifically, the presentation was excellent and very little, if any, fault could be found in that regard. In three categories a very high standard was achieved but some omissions and oversights meant that they did not warrant excellent (i.e. first class) marks. The discussion appears to have been somewhat limited because the student inexplicably appears not to have adequately addressed the fifth objective stated in the introduction. Therefore, there are signs of relative strength and also of a weakness, but on the balance of the evidence, this dissertation warranted a mark in the range of 60–69 percent (upper second class). It appears that this examiner may have taken the view that the limitation associated with the objective was sufficiently important to justify a mark just under half way through that classification.

Thus, as Box 15.1 and Tables 15.1 and 15.2 demonstrate, the greater the level of consistency in your performance, the higher your level of achievement is likely to be. However, in parallel to consistency, *continuity* is also vital, but this is often overlooked by students (hence our advice in this book).

One of the simplest checks an examiner can make is to ensure that the objectives (and/or research questions) are consistently articulated and properly addressed in the text (i.e. in the introduction, the methods and conclusion chapters), or that the same story is being told at the beginning, middle and end, to use the metaphor of the novel. In the case of the former, it is poor practice if there are discrepancies in the wording throughout your dissertation. Changing the wording (even subtly) can have a dramatic effect on the direction the research should take. Thus, what an examiner might have expected to encounter after the introduction as a result of one set of objectives may be different to what is actually presented, because the objectives (or research questions) are not, for whatever reason, the same later in the dissertation.

Furthermore, in the case of the latter, it is poor practice if you articulate objectives in the introduction only to overlook them in the data collection, analysis and hence in the conclusion of the dissertation. If the idea of the introduction is to set out your intentions for the dissertation, then clearly by omitting to address objectives at some stage in your dissertation your work is flawed because you have failed to achieve what you set out to do. Imagine how the examiner might consider such disconnections (Table 15.1) which can be avoided if you concentrate on continuity.

Who will assess my dissertation work?

Again the supporting materials provided by your institution or your advisor/s will answer this question directly. In general, you are likely to find that your dissertation *text* will be assessed by at least two members of staff. Your advisor/s may be also among your examiners, although some institutions frown on this practice because they consider that advisor/s are too close to your work and hence not able to make the most objective judgements of it.

Assessment by at least two examiners is good practice, not least because there may be so many credits riding on your dissertation. Your institution will want to be sure that this pinnacle of your studies is accurately assessed and that its students are graduating with the results that most appropriately reflect their abilities. However, don't worry if there is only one examiner. The total number of students in your programme may make it impossible to have double-marking from a practical perspective and/or it may simply not be your institution's policy to require double-marking for undergraduate dissertations.

This is unlikely, though. If your work is being assessed by two academics, it is important to establish what their roles are. There are two common scenarios. As you may be aware, there is a difference between marking and moderation. First and less common, the first examiner will read your work, assess it and compile (feedback) comments. As a quality assurance measure a second (internal) examiner will look at a sample of dissertations – perhaps but not necessarily yours – in order to moderate the marking. This is to ensure consistency in the marks being awarded versus the marking scheme, between students and between assessors. This is a more common practice for your taught modules.

Second and more routine is that two (internal) examiners will read your work independently, their assessments and comments will be compared, and a final (agreed) overall grading will be made. In some cases, the second examiner may be aware of the first examiner's comments, perhaps reading your dissertation after the first examiner and commenting on the same assessment pro forma. In others, a so-called 'double blind marking' system is used in which the examiners are not aware of each other's grade and comments when they assess your work. Once they have both completed their evaluations, they will discuss your work and decide on a final mark. If there is a significant discrepancy in their views, a third (internal) examiner is usually asked to assess your work 'blind' (i.e. without recourse to the first two examiners' comments). If there is still no agreement internally (and perhaps even if there is, as a matter of policy), your dissertation may be referred to an external examiner; that is, an academic from another institution whose role it is assure the quality of your programme and the work being conducted as part of it. Normally, the external examiner's role is to moderate student work but s/he may be called upon to make final judgements or confirm internal decisions on dissertations where internal agreement has previously proved problematic.

Life beyond the dissertation

In many respects, it may be natural to think of the completion of your dissertation and its examination as a point of closure in your life. You have your results, you know what the examiners thought of your dissertation, but what do you do next?

Many students will go on to get jobs and your dissertation should enhance your employability. In conducting your independent research you will have developed many transferable skills which will be of interest to future employers, including your literacy, numeracy, self-motivation, time management, project management, self-appraisal, liaison and problem-solving to name but a few.

Who knows, the dissertation may have even sparked your interest to conduct further study at master's and/or at doctoral level. After all, many degrees at master's level contain a dissertation and at doctoral level the preparation and successful defence of a substantial dissertation is *the* requirement for the degree. If so, we hope you'll continue to find this book helpful and good luck whatever you chose to do.

The chapter at a glance

The main learning points of this chapter are that:

- **Your dissertation work does not necessarily end the moment you submit your text.**
- **There are several different models of assessment for dissertation modules and courses.**
- **Assessment schemes are more nuanced and flexible than you might think.**
- **Several people are likely to read your work after it has been submitted.**
- **Continuity as well as consistency is vital to a successful dissertation.**

Dissertation checklist

In coming to the end of your dissertation experience, check you:

1.	Have read your institution's guidance on how your dissertation will be assessed.	
2.	Know who will read your work and why.	
3.	Understand how the marking scheme and assessment criteria work.	

REFERENCES

ABS (Association of Business Schools) (2011) *ABS Academic Journal Quality Guide*. Version 4, 2011. Online document. Available from: www.associationofbusinessschools.org/node/1000257 (last accessed: 14/08/2012).

Aguinis, H. and Harden, E.E. (2009) 'Sample size rules of thumb: evaluating three common practices', in Lance, C.E. and Vandenburg, R.J. (eds.) *Statistical and Methodological Myths and Urban Legends: Doctrine, Verity and Fable in the Organizational and Social Sciences*. New York: Routledge, 269–88.

Aguinis, H., Werner, S., Lanza Abbott, J., Angert, C., Park, J.H. and Kohlhausen, D. (2010) 'Customer-centric science: reporting significant research results with rigor, relevance and practical impact in mind', *Organizational Research Methods*, 13(3): 515–39.

Alvesson, M. (2011) *Interpreting Interviews*. London: Sage.

Alvesson, M. and Sköldberg, K. (2009) *Reflexive Methodology: New Vistas for Qualitative Research*. London: Sage.

Baggio, R. and Klobas, J. (2011) *Quantitative Methods in Tourism: A Handbook*. Bristol: Channel View.

Baranowski, S. and Furlough, E. (eds.) (2001) *Being Elsewhere: Tourism, Consumer Culture and Identity in Modern Europe and North America*. Ann Arbor, MI: University of Michigan Press.

Barbour, R. (2008) *Introducing Qualitative Research: A Student Guide to the Craft of Doing Qualitative Research*. London: Sage.

Barron, P. and Arcodia, C. (2002) 'Linking learning style preferences and ethnicity: international students studying hospitality and tourism management in Australia', *Journal of Hospitality, Sport and Tourism Education*, 1(2): 15–27.

Bazeley, P. (2007) *Qualitative Data Analysis with NVIVO*. London: Sage.

Bengry-Hall, A., Wiles, R., Wild, M. and Crow, G. (2011) *A Review of the Academic Impact of Three Methodological Innovations: Netnography, Child-led Learning and Creative Research Methods*. University of Southampton: ESRC National Centre for Research Methods, NCRM Working Paper Series 01/11.

Bergman, M. (2011) 'The politics, fashions and conventions of research methods', *Journal of Mixed Methods Research*, 5(2): 99–102.

Blaxter, L., Hughes, C. and Tight, M. (1996) *How to Research*. Buckingham Open University Press.

Bloor, M., Frankland, J., Thomas, M. and Robson, K. (2001) *Focus Groups in Social Research*. London: Sage.

Blumberg, K. (2008) 'Internationalisation in adventure tourism: the mobility of people, products and innovations', in Coles, T.E. and Hall, C.M. (eds.) *International Business and Tourism: Global Issues, Contemporary Interactions*. London: Routledge, 167–80.

Bochaton, A. and Lefebvre, B. (2011) 'Interviewing elites: perspectives from the medical tourism sector in India and Thailand', in Hall, C.M. (ed.) *Fieldwork in Tourism: Methods, Issues and Reflections*. London: Routledge, 70–80.

Booth, W.C., Gregory, G.C., Williams, J.M (2003) *The Craft of Research* (2nd ed.). Chicago, IL: University of Chicago Press.

Brotherton, B. (2008) *Researching Hospitality and Tourism: A Student Guide.* London: Sage.

Bryman, A. (2004) *Social Research Methods* (2nd ed.). Oxford: Oxford University Press.

—— (2007) 'Barriers to integrating quantitative and qualitative research', *Journal of Mixed Methods Research*, 1(1): 8–22.

Bryman, A. and Cramer, D. (2011) *Quantitative Data Analysis with IBM SPSS 17, 18 and 19: A Guide for Social Scientists.* London: Routledge.

Buglear, J. (2010) *Stats Means Business: Statistics with Excel for Business, Hospitality and Tourism* (2nd ed.). Oxford: Butterworth-Heinemann.

Burke, R. (2003) *Project Management: Planning and Control Techniques* (4th ed.). Chichester: Wiley.

Butler, R.W. (1980) 'The concept of a tourist area lifecycle of evolution: implications for the management of resources', *Canadian Geographer*, 24(1): 5–12.

—— (ed.) (2006) *The Tourism Lifecycle* (2 volumes). Clevedon: Channel View.

Cadman, K. (1997) 'Thesis writing for international students: a question of identity?' *English for Specific Purposes* 16 (1): 3–14.

Charmaz, K. (2004) 'Grounded theory', in Lewis-Beck, M.S., Bryman, A. and Liao, T.F. (eds) *The Sage Encyclopaedia of Social Science Research Methods (Volumes 1–3).* Thousand Oaks, CA: Sage.

Clark, T. (2010) 'On "being researched": why do people engage with qualitative research?', *Qualitative Research*, 10(4): 399–419.

Clewes, D. (1996) 'Multiple perspectives on the undergraduate project experience', *Innovations – The Learning and Teaching Journal of Nottingham Trent University*: 27–35.

Coles, T.E. (2004) 'Tourism and leisure: reading geographies, producing knowledges', *Tourism Geographies*, 6(2): 1–8.

—— (2008) 'Citizenship and the state: hidden features in the internationalisation of tourism', in Coles, T.E. and Hall, C.M. (eds) *International Business and Tourism: Global Issues, Contemporary Interactions.* London: Routledge, 55–69.

Coles, T.E. and Hall, D.R. (2005) 'Tourism and EU enlargement: plus ça change', *International Journal of Tourism Research*, 7(2): 51–62.

Coles, T.E., Hall, C.M. and Duval, D.T. (2006) 'Tourism and post-disciplinary enquiry', *Current Issues in Tourism*, 9 (4/5): 293–319.

Coles, T.E., Liasidou, S. and Shaw, G. (2009) 'Tourism and new economic geography: issues and challenges in moving from advocacy to adoption', *Journal of Travel and Tourism Marketing*, 25(3/4): 312–24.

Collis, J. and Hussey, R. (2003) *Business Research: A Practical Guide for Undergraduate and Postgraduate Students.* Oxford: Palgrave.

Cooper, H. (2010) *Research Synthesis and Meta-Analysis: A Step-by-Step Approach* (4th edition). Thousand Oaks, CA: Sage.

Corbin, J. and Strauss, A. (2008) *Basics of Qualitative Research* (3rd ed.). Los Angeles, CA: Sage.

Creswell, J.W. (1998; 2006; 2012) *Qualitative Inquiry and Research Design: Choosing Among Five Approaches* (1st, 2nd, 3rd eds). London: Sage.

—— (2003) *Research Design: Qualitative, Quantitative and Mixed Methods* (3rd ed.). London: Sage.

Darcy, S. (1998) *Anxiety to Access: Tourism Patterns and Experiences of New South Wales People with a Physical Disability.* Sydney: Tourism New South Wales.

Decrop, A. (1999) 'Triangulation in qualitative tourism research', *Tourism Management*, 20(1): 157–61.

Del Casino, V., Grimes, A.J., Hanna, S.P. and Jones, J.P. (2000) 'Methodological frameworks for the geography of organizations', *Geoforum*, 31(4): 523–38.

Denscombe, M. (2007) *The Good Research Guide: For Small-scale Social Research* (3rd ed.). Buckingham: Open University Press.

Dillman, D.A. (2007) *Mail and Internet Surveys: The Tailored Design Method* (2nd ed.). New York: J. Wiley and Sons.

Duval, D.T. (2004) '"When buying into the business, we knew it was seasonal": perceptions of seasonality in Central Otago, New Zealand', *International Journal of Tourism Research*, 6(5): 325–37.

—— (2008) 'Claim you are from Canada, eh: traveling citizenship within global space', in Burns, P. and Novelli, M. (eds) *Tourism and Mobilities: Local-Global Connections*. Wallingford: CABI, 81–91.

Exley and O'Malley (1999) 'Supervising PhDs in science and engineering', in Wisker, G. and Sutcliffe, N. (eds.) *Good Practice in Postgraduate Supervision*. London: Staff and Educational Development Association, SEDA Paper 106, 39–56.

Feilzer, M.Y. (2009) 'Doing mixed methods research pragmatically: implications for the rediscovery of pragmatism as a research paradigm', *Journal of Mixed Methods Research*, 4(1): 6–16.

Field, A. (2009) *Discovering Statistics Using SPSS* (3rd ed.). London: Sage.

Finn, M., Elliot-White, M. and Walton, M. (2000) *Research Methods for Leisure and Tourism: Data Collection, Analysis and Interpretation*. Harlow: Longman.

Franklin, A. and Crang, M. (2001) 'The trouble with tourism and travel theory', *Tourist Studies*, 1(1): 5–22.

Gallarza, M.G., Saura, I.G. and Garcia, H.C. (2002) 'Destination image: towards a conceptual framework', *Annals of Tourism Research* 29(1): 56–78.

Gelman, A. and Stern, H. (2006) 'The difference between "significant" and "not significant" is not itself statistically significant', *The American Statistician*, 60(4): 1–4.

Geertz, C. (1973) *The Interpretation of Cultures*. New York: Basic Books.

Hall, C.M. (2010) 'Crisis events in tourism: subjects of crisis in tourism', *Current Issues in Tourism*, 13(5): 401–17.

—— (ed.) (2011a) *Fieldwork in Tourism: Methods, Issues and Reflections*. London: Routledge.

—— (2011b) 'Biosecurity, tourism and mobility: institutional arrangements for managing tourism-related biological invasions', *Journal of Policy Research in Tourism, Leisure and Events*, 3(3): 256–80.

—— (2011c) 'Seeing the forest for the trees: tourism and the International Year of Forests', *Journal of Heritage Tourism*, 6(4): 271–83.

Hall, C.M. and Coles, T.E. (2008) 'Introduction: tourism and international business – tourism as international business', in Coles, T.E. and Hall, C.M. (eds) *International Business and Tourism: Global Issues, Contemporary Interactions*. London: Routledge, 1–26.

Hall, C.M. and Sharples, L. (2003) 'The consumption of experiences or the experience of consumption? An introduction to the tourism of taste', in Hall, C.M., Sharples, L., Mitchell, R., Macionis, N. and Cambourne, B. (eds) *Food Tourism around the World: Development, Management and Markets*. Oxford: Butterworth-Heinemann, 1–24.

Hall, C.M. and Valentin, A. (2005) 'Content analysis', in Ritchie, B.W., Burns, P. and Palmer C. (eds) *Tourism Research Methods: Integrating Theory with Practice*. Wallingford: CABI: 191–209

Hall, C.M. and Williams, A.M. (2008) *Tourism and Innovation*. London: Routledge.

Hart, C. (2005) *Doing your Masters Dissertation*. London: Sage.

Harvey, W.S. (2011) 'Strategies for conducting elite interviews', *Qualitative Research*, 11(4): 431–41.

Healey, M. and Healey, R.L. (2003) 'How to conduct a literature search', in Clifford, N. and Valentine, G. (eds) *Research Methods in Human and Physical Geography*. Sage: London: 16–31.

Hickman, L. (2007) *The Final Call*. Cornwall: Eden Project.

Hine, C. (2000) *Virtual Ethnography*. London: Sage.

Holden, A. (2005) *Tourism Studies and the Social Sciences*. London: Routledge.

Holloway, J.C. and Robinson, C. (1995) *Marketing for Tourism* (3rd ed.). Harlow: Longman.

Huang, R. (2007) 'A challenging but worthwhile learning experience: Asian international student perspectives on undertaking a dissertation in the UK', *Journal of Hospitality, Leisure, Sport and Tourism Education*, 6(1): 29–38.

Hussey, J. and Hussey, R. (1997) *Business Research: A Practical Guide for Undergraduate and Postgraduate Students*. London: Macmillan Press.

I'Anson, R.A. and Smith, K.A. (2004) 'Undergraduate research projects and dissertations: issues of topic selection and data collection amongst tourism management students', *Journal of Hospitality, Leisure, Sport and Tourism Education*, 3(1): 19–32.

Israel, G.D. (1992) *Determining Sample Size*. University of Florida: Program Evaluation and Organizational Development, Florida Co-operative Extension Service, Institute of Food and Agricultural Services (Fact Sheet PEOD-6).

Jack, G. and Phipps, A. (2005) *Tourism and Intercultural Exchange: Why Tourism Matters*. Clevedon: Channel View.

Jafari, J. (ed.) (2000) *Encyclopaedia of Tourism*. London: Routledge.

Johnson, R.B., Onwuegbuzie, A.J. and Turner, L.A. (2007) 'Toward a definition of mixed methods research', *Journal of Mixed Methods Research*, 1(2): 112–33.

Jones, N. (2004) *Rough Guide Special: Travel Health. Planning your Trip Worldwide*. London: Rough Guides/Penguin.

Kaplowitz, M.D., Hadlock, T.D. and Levine, R. (2004) 'A comparison of web and mail survey response rates', *Public Opinion Quarterly*, 68(1): 94–101.

Knapp, T.R. and Mueller, R.O. (2010) 'Reliability and validity of instruments', in Hancock, G.R. and Mueller, R.O. (eds) *The Reviewer's Guide to Quantitative Methods in the Social Sciences*. London: Routledge, 337–42.

Kozinets, R.V. (2002). 'The field behind the screen: using netnography for marketing research in online communities', *Journal of Marketing Research*, 39(1), 61–72.

Krejcie, R.V. and Morgan, D.W. (1970) 'Determining the sample size for research activities', *Educational and Psychological Measurement*, 30: 607–10.

Lance, C.E. (2011) 'More statistical and methodological myths and urban legends', *Organizational Research Methods*, 14(2): 279–86.

Law, J. (2004) *After Method: Mess in Social Science Research*. London: Routledge.

Lee, S. and Park, S-Y. (2009) 'Do socially responsible activities help hotels and casinos achieve their financial goals?', *International Journal of Hospitality Management*, 28(1): 105–12.

Lew, A.A., Hall, C.M. and Williams, A.M. (eds) (2004) *A Companion to Tourism*. Oxford: Blackwell.

Lerner, R.M. (2002) *Concepts and Theories of Human Development* (3rd ed.). Mahwah, NJ: Lawrence Erlbaum Associates.

Lowes, R., Peters, H. and Turner, M. (2004) *The International Student's Guide to Studying English at University*. London: Sage.

Markusen, A. (2003) 'Fuzzy concepts, scanty evidence, policy distance: the case for rigour and policy relevance in critical regional studies', *Regional Studies*, 37(6/7): 701–17.

Morgan, N.J. and Pritchard, A.J. (1999) *Power and Politics at the Seaside: The Development of Devon's Resorts in the Twentieth Century*. Exeter: University of Exeter Press.

Mosedale, J.T. (2007) 'Corporate Geographies of Transnational Tourism Corporations' unpublished PhD Thesis, University of Exeter.

Munar, A.M. (2007) 'Is the Bologna process globalizing tourism education?', *Journal of Hospitality, Leisure, Sport and Tourism Education*, 6(2): 68–82.

Murphy, P. and Murphy, A. (2004) *Strategic Management for Tourism Communities: Bridging the Gaps*. Clevedon: Channel View.

New Jour (2012) *New Jour Search Results*. Results for tourism 1 to 15 of 44 results. Online document. Available from: http://gulib.georgetown.edu/newjour/search/do?query=tourism &metaname=swishtitle&submit=Search (last accessed: 14/08/2012).

O'Leary, Z. (2010) *The Essential Guide to Doing Your Research Project*. London: Sage.

Oliver, P. (2008) *Writing Your Thesis* (2nd ed.). London: Sage.

Page, S.J. and Connell, J. (2011) *The Routledge Handbook of Events*. London: Routledge.

Parsons, T. and Knight, P.G. (2005) *How to Do Your Dissertation in Geography and Related Subjects* (2nd ed.). London: Routledge.

Phillimore, J. and Goodson, L. (2004) *Qualitative Research in Tourism: Ontologies, Epistemologies and Methodologies*. London: Routledge.

Phillips, E.M. and Pugh, D.S. (1994) *How to Get a PhD: A Handbook for Students and their Supervisors*. Buckingham: Open University Press.

—— (2000) *How to Get a PhD: A Handbook for Students and their Supervisors* (3rd ed.). Buckingham: Open University Press.

Rakic, T. and Chambers, D. (2011) *An Introduction to Visual Methods in Tourism*. London: Routledge.

Rapley, T. (2011) 'Some pragmatics of qualitative data analysis', in Silverman, D. (ed.) (2011) *Qualitative Research* (3rd ed.). London: Sage, 273–90.

Reisinger, Y. and Turner, L. (2003) *Cross-cultural Behaviour in Tourism: Concepts and Analysis*. Oxford: Butterworth-Heinemann.

Ridley, D. (2008) *The Literature Review: A Step-by-step Guide for Students*. London: Sage.

Ritchie, B.W. (2009) *Crisis and Disaster Management for Tourism*. Clevedon: Channel View.

Ritchie, B.W., Burns, P. and Palmer, C. (eds) (2005) *Tourism Research Methods: Integrating Theory with Practice*. Wallingford: CAB International.

Rossman, G. and Rallis, S. (1999) *Designing Qualitative Research* (3rd ed.). Thousand Oaks, CA: Sage.

Ryan, C. (1995) *Researching Tourist Satisfaction: Issues, Concepts and Problems*. London: Routledge.

Salt, C.M. (2004) *Interactive Barriers and Visual Representation: A Study of Tourists with Disabilities*. Unpublished MSc Dissertation, University of Exeter.

Sandelowski, M. and Barroso, J. (2006) *Handbook for Synthesizing Qualitative Research*. New York: Springer.

Saunders, M.N.K., Lewis, P. and Thornhill, A. (2000) *Research Methods for Business Students* (2nd ed.). London: Pitman Publishing.

Scherle, N. and Coles, T.E. (2008) 'International business networks and intercultural communications in the production of tourism', in Coles, T.E. and Hall, C.M. (eds) *International Business and Tourism: Global Issues, Contemporary Interactions*. London: Routledge, 124–42.

Scott, J. (1990) *A Matter of Record: Documentary Sources in Social Research*. Cambridge: Polity.

Semmens, K. (2005) *Seeing Hitler's Germany: Tourism in the Third Reich*. Basingstoke: Palgrave Macmillan.

Shaw, G. and Williams, A.M. (2002) *Critical Issues in Tourism* (2nd ed.). Oxford: Blackwell.

—— (2004) *Tourism and Tourism Spaces*. London: Sage.

—— (2009) 'Knowledge transfer and management in tourism: an emerging research agenda', *Tourism Management*, 30: 325–35.

Silverman, D. (ed.) (2011) *Qualitative Research* (3rd ed.). London: Sage.

Sönmez, S. (1998) 'Tourism, terrorism and political instability', *Annals of Tourism Research*, 25(2): 416–56.

Stapleton, L. (2010) 'Survey sampling, administration and analysis', in Hancock, G.R. and Mueller, R.O. (eds) *The Reviewer's Guide to Quantitative Methods in the Social Sciences*. New York: Routledge, 397–412.

Stephenson, M. (2002) 'Travelling to the ancestral homeland: the aspirations and experiences of a UK Caribbean community', *Current Issues in Tourism*, 5(5): 378–425.

Stern, N. (2007) *The Economics of Climate Change: The Stern Review*. Cambridge: Cambridge University Press.

Swarbrooke, J. (2003) 'Corporate social responsibility and the UK tourism industry', *Insights, Tourism Intelligence Papers*, A75-A83.

Tashakkori, A. and Creswell, J. (2007) 'Editorial: the new era of mixed methods', *Journal of Mixed Methods Research*, 1(1): 3–7.

Tarling, R. (2006) *Managing Social Research: A Practical Guide*. London: Routledge.

Thomas, R., Shaw, G. and Page, S.J. (2011) 'Understanding small firms in tourism: a perspective on research trends and challenges', *Tourism Management*, 32(5): 963–76.

—— (2009) (ed.) *Philosophical Issues in Tourism*. Clevedon: Channel View.

Veal, A.J. (2011) *Research Methods for Leisure and Tourism: A Practical Guide* (4th ed.). Harlow: Prentice Hall/Financial Times.

Veitch, C. and Shaw, G. (2004) 'Understanding barriers to tourism in the UK', *Insights, Tourism Intelligence Papers*, A185-A193.

VisitBritain (2003) *Holiday-taking and Planning amongst People with a Disability*. London: NOP.

Weaver, D.B. (ed.) (2003) *The Encyclopaedia of Ecotourism*. Wallingford: CAB International.

Webster, F., Pepper, D. and Jenkins, A. (2000) 'Assessing the undergraduate dissertation', *Assessment and Evaluation in Higher Education*, 25(1): 71–80.

Wild, E. (2005) *The Influence and Effects of Film-Induced Tourism in Cornwall*. Unpublished MSc Dissertation, University of Exeter.

Williams, S.V. (ed.) (2003) *Tourism: Critical Concepts in the Social Sciences*. London: Routledge.

Woodhouse, M. (2002) 'Supervising dissertation projects: expectations of supervisors and students', *Innovations in Education and Teaching*, 39(2): 137–47.

WTTC (World Travel and Tourism Council) (2010) *Progress and Priorities, 2009–10*. London: WTTC.

—— (2011a) *Travel & Tourism 2011*. Online document. Available from: www.wttc.org/site _media/uploads/downloads/traveltourism2011.pdf (last accessed: 24/03/2012).

—— (2011b) *Economic Impact of Travel and Tourism*. Update November 2011. Online document. Available from: www.wttc.org/site_media/uploads/downloads/4pp_document_ for_WTM_RGB.pdf (last accessed: 24/03/2012).

—— (2012) *Travel & Tourism Economic Impact 2012 World*. Online document. Available from: www.wttc.org/research/economic-impact-research/ (last accessed: 03/02/2012).

INDEX

References to figures are shown in *italics*. References to tables are shown in **bold**. References to boxes are shown in ***bold italics***.